S0-ARK-258

A Birder's Guide to the Chicago Region

Joel Greenberg

A
Birder's
Guide
TO THE
Chicago
Region

Lynne Carpenter and Joel Greenberg

Introduction by Kenn Kaufman

NORTHERN

ILLINOIS

UNIVERSITY

PRESS

1999

© 1999 by Northern Illinois University Press

Published by the Northern Illinois University Press,

DeKalb, Illinois   60115

Manufactured in the United States using

acid-free paper

All Rights Reserved

Design by Julia Fauci

Library of Congress Cataloging-in-Publication Data

Carpenter, Lynne.

A birder's guide to the Chicago Region / Lynne

Carpenter and Joel Greenberg.

    p.   cm.

Includes bibliographical references.

ISBN 0-87580-582-5 (alk. paper)

1. Bird watching—Illinois—Chicago Region

Guidebooks. I. Greenberg, Joel (Joel R.) II. Title.

QL684.I3 C37 2000

598′ .07′23477311—dc21

99-35856

CIP

Cover photograph:

© James P. Rowan Photography

To the Evanston North Shore Bird Club,

which has provided inspiration and comraderie

to birders since 1919.

There is a natural bond of affinity between those who have gone to the fields in March to hear the tinkle of the horned larks on the frozen ground; or in the hope of seeing a meadow lark in the snowy stubble; or a group of early blue birds on fence or wire. There is deep understanding among those who have heard the mystical nuptial flight song of the woodcock at an April dusk; who have been surprised by the bubbling rapture of the ruby-crown's mating song in the budding woods of early May; who have listened with moist eyes to the robins singing in the rain, with its nostalgic memories of childhood; or who have waited in a reverent and breathless silence, while the wood thrush chimed his bells in a June twilight.

(Excerpt from the letter dated February 3, 1940, written by Fred Pattee to the Evanston North Shore Bird Club after the death of his wife, Bertha, the club's founder and president from 1919–1939.)

Contents

List of Maps   xi
Foreword by Kenn Kaufman   xiii
Preface   xv
Introduction   xvii

BIRDING IN THE CHICAGO REGION   3

Monthly Suggestions   5

BIRDING LOCATIONS   15

Illinois Sites

Chicago and Central Cook County   17
Northern Cook County   28
Southern Cook County   45
Lake County   65
McHenry County   92
DuPage County   102
Kane County   116
Kendall County   124
Grundy County   127
Will County   130
Kankakee County   144

Wisconsin Sites

Kenosha County   149
Racine County   152
Walworth County   155

Indiana Sites

Lake County   161
Porter County   166
LaPorte County   173
Newton County   177
Jasper County   180
Starke County   182

Michigan Sites

Berrien County   185

BIRDS OF THE CHICAGO REGION   *193*

Appendices
    A: Resources   215
    B: Nature Centers   219
    C: Hotlines   223
    D: Christmas Bird Counts   225
    E: Rare Bird Documentation   227

Bibliography   229
Index to Species   231
Index to Sites   237

Maps

Natural Divisions of the Region of Chicago   *xviii*
Chicago Region   *2*

Illinois Maps

Chicago and Central Cook County (M)   *16*
Magic Hedge and Montrose Harbor   *19*
Jackson Park   *22*
Northeastern Cook County (K)   *29*
Skokie Lagoons   *34*
Chicago Botanic Garden   *35*
Northwestern Cook County (J)   *39*
Southwestern Cook County (P)   *45*
Palos Preserves   *46*
Southern Cook County (Q)   *53*
Lake Calumet area   56
Southeastern Lake County (G)   *66*
Ryerson Woods   *70*
Northeastern Lake County (E)   *73*
Waukegan Harbor Area   *74*
Illinois Beach State Park   *76*
Southwestern Lake County (F)   *80*
Northwestern Lake County (D)   *84*
Chain O'Lakes State Park   *91*
McHenry County (C)   *93*
Glacial Park Conservation Area   *94*
Moraine Hills State Park   *99*
DuPage County (L)   *101*
Fermilab   *104*
Morton Arboretum   *110*
Kane County (H)   *117*
Kendall and Grundy Counties (N)   *125*
Goose Lake Prairie Area   *128*
Western Will and Kankakee Counties (R)   *131*
Northern Will County (O)   *136*
Eastern Will and Kankakee Counties (S)   *142*

Wisconsin Maps

Kenosha and Racine Counties (B)   *148*
Walworth County (A)   *156*

Indiana Maps

Northwestern Indiana (V)   *160*
North Central Indiana (W)   *167*
Indiana Dunes State Park   *171*
Greater Indiana (T)   *175*
Kingsbury Fish and Wildlife Area   *176*
Willow Slough Fish and Wildlife Area   *179*
Jasper-Pulaski Fish and Wildlife Area   *181*

Michigan Maps

Berrien County (X)   *184*
Warren Dunes State Park   *188*

Foreword

Kenn Kaufman

ALTHOUGH IT SEEMS A PARADOX, some of the best birdwatching locations in North America are not in remote wilderness areas but on the outskirts
of major cities. I can think of no better example of this than the number of fine
birding sites that surround the great city of Chicago.

My own introduction to this region came in June 1972, when I went to visit
Joel Greenberg at his family's home northwest of the city. Although we were
just teenagers, Joel already knew the birds of that area extremely well; in a few
days of blasting around the metro area we saw a remarkable number of species,
from Cerulean Warbler (typical of southeastern forest) to Brewer's Blackbird
(typical of the wide-open west). Most interesting of all was the Henslow's Sparrow, a flat-headed, flat-voiced little gnome, that Joel found for me at Goose
Lake Prairie. He was able to find it easily because he understood not only the
bird but also the details of its habitat, presaging his later work in broader aspects of natural history.

In later years I learned that the Chicago region offered exciting birds at all
seasons. Winter brought birds down from the far north, along with scattered
strays from far afield. Spring and fall brought great concentrations of migrants,
especially along the lakeshore. Throughout the year, a corps of active birders
was out combing the woods and fields and watersides, determining the best
spots to find every species. The birders' version of the Chicago region—encompassing nineteen counties in four states, wrapping around the lakeshore from
southeastern Wisconsin to southwestern Michigan—is surely one of the best-
known areas, ornithologically, in North America.

Despite the abundance of birdlife and bird knowledge in this region, there
has never been a really comprehensive guide to bird-finding there. One person
who recognized that lack was Lynne Carpenter, an active area birder and former
president of the Evanston North Shore Bird Club. Lynne began compiling detailed information on birding sites, based on her own extensive experience and
on input from many birders throughout the four-state region. Joel Greenberg
joined the project, bringing to bear his expertise on birds and other facets of nature. These two authors have succeeded in producing a guide that packs a
wealth of helpful information into a compact, easy-to-use book.

This volume guides us to more than 250 prime birding locations in the 19-
county area surrounding Chicago. The detailed directions (and fine accompany-
ing maps) not only get us to the sites, but tell us what to do when we get there,
with information on parking, trails, and other access. The directions alone would
be enough to make this book essential. But there is more here, too. The list of
month-by-month birding suggestions contains a host of ideas to get birders out

into the field at every season. The annotated list of species lets us know which times of year to expect each bird, and which habitats they favor. One thing I particularly like is the inclusion of numerous notes on natural history and background—allowing us to understand WHY these sites offer productive birding. In this era, as more and more birdwatchers develop an interest in wider aspects of nature, this kind of interpretation is especially welcome.

This will be an indispensable guide for anyone with an interest in birds who either lives in the greater Chicago region or has the opportunity to visit there. I recommend it highly, and congratulate the authors on helping to make the birdlife of their region more accessible and more interesting for all of us.

Preface

MOST PEOPLE ARE AMAZED TO LEARN that over 300 species of birds are seen in the Chicago area each year. The average resident, when pressed, may be able to name such common birds as Canada Goose, American Crow, Robin, Nothern Cardinal or House Sparrow. Those with feeders or bird baths may have attracted additional species to their yards, and others may have noticed a wider variety while engaging in outdoor activities such as hiking, fishing or boating. Sometimes, a teacher, relative or neighbor provides the spark that ignites a curiosity about birds. When that curiosity strikes, the purchase of a book, a pair of binoculars or an inquiry into the activities of a local nature center or club may mark the beginning of a lifelong interest in birds.

This book is a guide intended to assist both beginners and veterans in seeing more birds in the Chicago area. Over 250 sites from across the region are described. Previous guides to bird finding in Chicago are out of date, out of print, or describe only a portion of the Chicago area. Any list of good birding locations is apt to change over time. Change occurs when sites are lost or impaired by development, but positive changes occur, too, as sites become accessible after being acquired by public or private conservation agencies. By practicing sound land management, these agencies can then maintain or enhance the habitat to support birds. A number of wetland creation projects in the Chicago area, for example, have successfully attracted migrant and breeding species dependent upon such ecosystems.

The great task is to ensure that there will always be places in the Chicago area to accommodate the birds and other wildlife that mean so much to mankind. By showcasing the avian diversity of this region, this book hopes to contribute to the successful meeting of that challenge.

For this guide, birders in all parts of the Chicago area submitted information about their favorite birding sites and the birds they look for. Ken Brock permitted information from his definitive "Birds of the Indiana Dunes" to be included, as did contributors to "Wisconsin's Favorite Bird Haunts" compiled and edited by Daryl Tessen. Scott Hickman, David Johnson, Andy Sigler and Al Stokie were especially generous in sharing their expertise. Other contributors include Cindy Alberico, Alan Anderson, Renee Baade, Bill Banks, Joy Bower, John Bielefeldt, Laurence Binford, Richard Biss, Thon Boves, David Brooks, Nan Burkhart, Jim Carpenter, Mike Carpenter, Paul Clyne, Hal Cohen, John Csoka, Helen Dancey, Donald Dann, Jerry De Boer, Dennis De Courcey, Sheryl DeVore, Danny Diaz, Peter Dring, Aura Duke, John Elliott, Carol and Conrad Fialkowski, Carolyn Fields, Karen and Bob Fisher, Darlene Fiske, Ron Flemal, Sue Friscia, Ray Ganey, Kirk Garanflo, Urs Geiser, Bill Glass, Marianne Hahn, Duane Heaton, Mary Hennen, Ralph Herbst, Libby Hill, Mel Hoff, Ron Hoffman, Roger Hotham, Richard Hugel, Robert Hughes, Mark Hurley, Doris Johansen, Denis Kania, Peter Kasper, Rose Kirwin, Vernon Kleen, Walter Krawiecz, Dennis Lane, Jim Landing, Christine and

Steven Lee, Francis Lenski, Joe Lill, Michel Madsen, David Mandell, Walter Marcisz, Carolyn Marsh, Judy Mellin, Margo Milde, David Miller, Kip Miller, Joe Milosevich, Bob Montgomery, William Moskoff, Stacia Novy, John O'Brien, Jerrold Olson, Pat Parsons, Wendy Paulsen, Dick Plank, Jack Pomatto, Chuck Roth, Jeffrey Sanders, Marv Schwartz, Valdemar Schwarz, Brad Semel, Wes Serafin, Darrell Shambaugh, Roberta Simonds, Audrey Smith, Muriel Smith, Mark Spreyer, Bobbie Squires, Jim Steffen, Doug Stotz, Paul Strand, Joe Suchecki, Mary Jane Sweeney, Floyd Swink, Anton Szabados, Craig Thayer, Ron Valani, Eric Walters, Chuck Wescott, David Willard, Dan Williams, Kenneth Wysocki, and Dick Young. The book would not have been possible without their contributions; we are grateful to all of them.

Introduction

The Chicago Region

THE GIANT ICE FLOW known as the Wisconsin Glacier created most of the topographic features of the Chicago Region. For thousands of years, its sporadic ebb and flow deposited vast quantities of rock, earth, and water. The distribution of these materials form seven natural divisions.

Three of the divisions are morainal uplands that parallel Lake Michigan in concentric "horseshoes." Going away from Lake Michigan, they are the Lake Border Upland, Valparaiso Upland, and Outer Upland. These moraines, called end or terminal moraines, formed at the glacier's edge, and represent massive depositions at a period when the glacier was stalled. The Valparaiso Upland, the highest of the three, features not only the most striking scenery but also hundreds of lakes, concentrated in the northwestern part of the region and, to a lesser extent, northern Indiana. Where the glacier receded without stopping, it left behind till plains, relatively flat areas marked by small rolling ground moraines. The Manteno Plain is such an area and is the smallest of the natural divisions.

As the glacier receded, meltwater formed lakes between the moraines and walls of ice. The Chicago Lake Plain and the Morris-Kankakee Basin are two natural divisions that were once the beds of these glacial lakes. These sections tend to have sandy soils and extensive wetlands.

The seventh natural division is Lake Michigan, dug by the glacier and filled with melt water. The lake, which is only about 2,000 years old, drains east toward the St. Lawrence River and the Atlantic Ocean. It was preceded by a series of lakes, whose shapes and drainage patterns changed as the glacier melted. The most significant of these was Lake Chicago. It was bounded by the Valparaiso Upland and the glacier, and it lasted for 4,000 years. Eventually, the chilly water surged southwesterly through a low point in the moraine, and scoured the lower Des Plaines River Valley from Willow Springs to Joliet as it headed to the Mississippi via the Illinois River. The rushing water exposed bedrock, upon which formed dolomite prairies that thrive in virtually no soil. Later, a second outflow developed to the south through an area known as the Sag Channel. Between the two branches, a triangular shaped upland remained. Known as Mount Forest Island, it is the heart of the Palos forest preserves.

The Valparaiso Moraine is one of the continent's principal divides. The Chicago River, Calumet River, Root River, Pike River, and St. Joseph River flowed through the Chicago Lake Plain and were in the Great Lakes Basin. (Engineering feats at the turn of the century shunted the Chicago River and the Calumet River away from Lake Michigan and into the Illinois River system.) All the other rivers in the region flow toward the Mississippi. The Des Plaines and Kankakee meet to form the Illinois River, the third largest tributary of the Mississippi. The

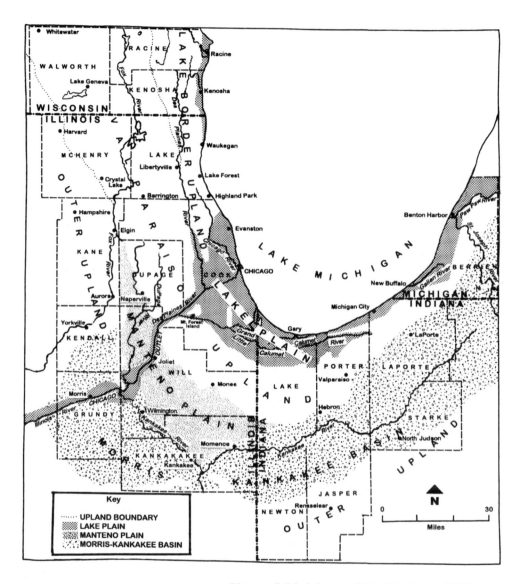

Natural Divisions of the Region of Chicago

Fox River flows south to meet the Illinois west of our region. Two small sections of the Chicago area are outside the Illinois River watershed: small portions of McHenry and Walworth Counties drain to the Rock River and the southern portions of Starke and Jasper Counties drain to the Ohio River.

Climate is the single most significant factor influencing vegetation. A cold period brought the ice and a warming period melted the ice. When the glacier eventually disappeared about 10,000 years ago, the region became dominated by northern forests of fir, spruce, and pine. Then, roughly 2,000 years later, a very warm period promoted grasslands, the tall grass prairies that are unique to the Midwest. Gradually, the shift has been toward cooler and moister times that favor trees.

The principal reason that forests did not totally reclaim this region is fire. It was of common occurrence in the old days, and fires were set by both lightning and people. Fire frequency, along with soil and moisture, determined whether land would be covered with prairie (dominated by grass), forest (dominated by trees), savanna (scattered trees with prairie understory) or barrens (hazel and other shrubs). And fire frequency itself was affected by topography. Prairies developed on level ground where fires pushed by predominantly westerly winds burned without obstructions. Forests survived in places where fires rarely reached, such as the east side of rivers and other water barriers, ravines, and behind high dunes. (Most of Berrien County was outside the prairie region.) Somewhere on the continuum between the extremes of topography and fire frequency, there were the transitional savannas and scrub barrens. It is also important to realize, although difficult today where only scraps of native landscapes survive, that the wetlands, prairies, woods, savannas, and barrens were part of a seamless whole, with borders that tended to blend and shift.

To better understand and protect the biological diversity of the Midwest, ecologists began to categorize different biological communities based primarily on vegetation. One such effort, the landmark Illinois Natural Areas Inventory (1978), determined that roughly 12 types of woods, 21 types of prairies, 4 types of savannas, and 17 types of wetlands occurred near Chicago. Some of these are referred to in the text in describing the special features of a given site.

From a strictly ornithological perspective, these distinctions are generally not of great importance. Birds, and other vertebrates, seem to respond to vegetational structure rather than specific composition. No bird currently in the region needs high quality native plant communities in which to nest. If a species were that restricted, it would have become extinct long ago. (At least one study, however, indicates that many insects do require high quality plant communities.) Prairie birds have adjusted to fields dominated by old world plants, and savanna birds thrive in woodland edges, orchards, or other areas of open trees. So long as there is an appropriate mix of vegetation and open water, most wetland birds will occur even if the vegetation is largely cattail and lacking the floral diversity characteristic of a pristine marsh. But if the structure is changed, such as opening up closed forests or allowing cattails to fill in open water, the avifauna will change as well.

The landscape of Chicago has been radically altered over the 300 years since Europeans first arrived. A tremendous amount has been lost, particularly when one thinks in terms of acreage. But what is remarkable is how much biological diversity remains. The sites listed in the following pages are testaments to that diversity.

Using This Guide

The "Chicago region" as defined in this book includes the 19 counties that lie within 65 miles of Chicago. Most (9) of the counties are in Illinois, 6 are in Indiana, 3 are in Wisconsin, and 1 is in Michigan. The book features the following sections:

Birding in the Chicago Region: This section discusses why birding in the region is unique and offers suggestions for what birds to look for and likely places to look for them in each month of the year.

W S S F

The above graphic shows both winter and spring to have a low score of one, summer to have a score of two, and fall to have a score of three.

The Sites: Over 250 birdwatching sites within 2 hours of downtown Chicago are plotted in regional maps. All sites are open to the public or visible from a public road or trail. The sites are described fully in the text with directions, trail information, telephone numbers, and other helpful information. The location of each site on its corresponding map follows the name of the site in parentheses. Sites are ranked by season (winter, spring, summer, fall) with a grid in the margin. Blackened squares signify a season's bird siting potential, with 4 black squares indicating the best season.

The birds listed for each site are some of the species that can be reasonably expected in a particular season. Because birds frequent different sites in different seasons, you should read the text to see if a site is attractive to the birds you want to see in the month you plan to visit. Lesser-known sites are included to encourage birders to explore them; rare and interesting birds may be found in these preserves as well.

Each map is drawn to scale, and all birding sites are numbered on the regional maps. The maps contain the major highways and roads birders need to get from one spot to another. For minor roads or other features, the use of commercial road maps such as the Chicago Tribune Map or the various state DeLorme Gazetteers is encouraged.

Birds of the Chicago Region: The species in this section, listed in the taxonomic sequence of the American Ornithologist's Union (AOU) seventh edition, 1998, represent the birds that occur in this region at least once every two years. It, therefore, omits accidental species. Periods of occurrence, abundance, and habitat are provided for each bird. Mention is also made of locations known to harbor species deemed to be of special interest to birders, either because of the species' rarity or localized distribution.

Appendices: Here the reader will find information about local birding organizations, Christmas Bird Counts, hotlines, and guidelines for documenting rare birds.

Cautionary Notes

Many birding sites are in isolated locations or in urban areas where the solo birder or his property may be at risk. For those unfamiliar with the area, it is prudent to ascertain the current status of a site or to bird with others. Use the same common sense necessary in any major U.S. city to ensure a safe outing.

Located as they are in one of the nation's largest metropolitan areas, the sites mentioned in this book are subject to high volumes of human traffic. Whether the area you are visiting is a high quality natural area or not, please follow the rules, remain on the trails, and tread gently. To avoid frustration or disappointment, call the site in advance to obtain current information on closings, special events, or construction.

A Birder's Guide to the Chicago Region

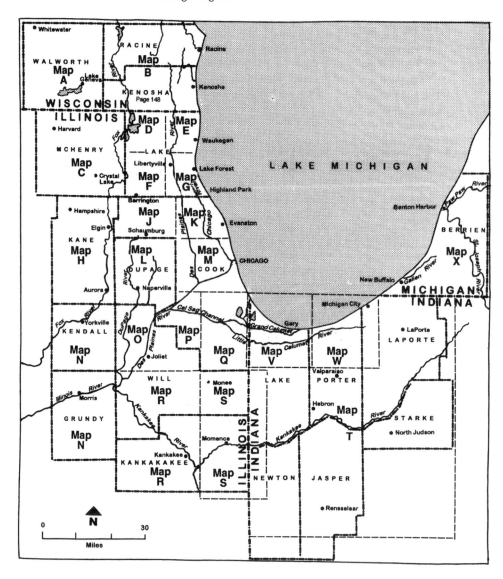

Chicago Region

A   Walworth County *p. 156*

B   Kenosha and Racine Counties *p. 148*

C   McHenry County *p. 93*

D   Northwestern Lake County *p. 84*

E   Northeastern Lake County *p. 73*

F   Southwestern Lake County *p. 80*

G   Southeastern Lake County *p. 66*

H   Kane County *p. 117*

J   Northwestern Cook County *p. 39*

K   Northeastern Cook County *p. 29*

L   DuPage County *p. 101*

M   Chicago and Central Cook County *p. 16*

N   Kendall and Grundy Counties *p. 125*

O   Northern Will County *p. 136*

P   Southwestern Cook County *p. 45*

Q   Southern Cook County *p. 53*

R   Western Will and Kankakee Counties *p. 131*

S   Eastern Will and Kankakee Counties *p. 142*

T   Greater Indiana *p. 175*

V   Northwestern Indiana *p. 160*

W   North Central Indiana *p. 167*

X   Berrien County *p. 184*

Birding in the Chicago Region

Thanks to its geography and wide variety of habitats, the Chicago region is host to millions of birds a year. Some are permanent residents, while others only stay to winter or to nest. Most, however, appear as migrants on their way between winter and summer destinations. Among these regulars may be found, sprinkled like treasured spices, the vagrant species whose presence adds so much to the excitement of local birding.

Fortunately, it is easy to experience this avian wealth. The average backyard has its familiar group of birds, augmented during migration by some that are perhaps not so familiar. But, to sample the full richness of the region's birdlife, one must visit the many parks, woods, grasslands, marshes, and lakes that grace the area.

This volume will make that endeavor more successful and rewarding. Knowing when and where to look will help to maximize the number of birds one encounters. For the reader who wishes to go beyond merely seeing birds, the book's appendices provide the information needed to become more involved. By becoming more active in the local birding community, the birder finds comraderie and contributes directly to the welfare of birds.

Monthly Suggestions

THE SAME WOODS that appear dormant in March might be swarming with warblers two months later; the lakes that yield only boats in summer may become spotted with waterfowl in April and May. The following suggestions are intended to give the birder an idea of what birds to look for each month and a selection of sites to search. Based upon the recommendations of birders with many years of experience, the suggestions are only a sampling of the possibilities. While the site names are often abbreviated here, the index contains full information.

January

Begin the month by taking part in one of the Chicago region's Christmas Bird Counts. The last dates for this annual census fall in the first week of January. Call your local bird organization to be assigned to an area. To learn if any unusual or rare birds have been seen, call the Rare Bird Alert hotline. In January, look for the following:

Waterfowl anywhere the water is open. In addition to Lake Michigan, try Lake Geneva, Heidecke Lake, Wolf Lake, Dresden Lake, the North Shore Channel and below the dams on the Fox River, the Des Plaines River, and the Illinois River;

Mute Swans on Wolf Lake;

Oldsquaws in Kenosha Harbor, Waukegan, Lake Forest, Ft. Sheridan, and Racine;

Red-Shouldered Hawks in wet woods including areas along the Des Plaines and Kankakee Rivers and in the Indiana Dunes;

Rough-legged Hawks at open areas such as Newton County, Ind., Fermilab, Midewin, Volo Bog, Spring Creek Valley, and Kane County;

Bald Eagles at Heidecke Lake;

Northern Harriers at Fermilab, Goose Lake Prairie, and Newton County;

Wintering gulls at Lake Calumet landfills, O'Brien Lock and Dam, and North Point Marina and along Lake Michigan or wherever there is open water. Gulls can be attracted by tossing bread on the water, an activity known as "chumming";

Great Horned Owls in oak woods at Orland Hills, River Trail, Brookfield Zoo, Port of Indiana, Mt. Baldy, and New Buffalo;

Snowy Owls at Meigs Field, Lake Calumet, Montrose Harbor, Navy Pier jetties, O'Hare, and New Buffalo;

Long-eared Owls in stands of pines at Orland Hills, Petrifying Springs, Fermilab, Chain O'Lakes, Deer Grove, Waterfall Glen, and Greene Valley;

Short-eared Owls at Bong State Recreation Area, Pratt's Wayne Woods, Midewin, Lake Calumet, Fermilab, and Kingsbury;

Northern Saw-whet Owls in yews and arbor vitae trees at Morton Arboretum, Petrifying Springs, Waterfall Glen, and Deer Grove;

Northern Shrikes at Volo Bog, Illinois Beach, Lyons Woods, Crabtree, Beverly Shores, and Kingsbury;

Winter finches at the Morton Arboretum, Crabtree, Illinois Beach, Indiana Dunes Nat. Lakeshore, Volo Bog, and Lyons Woods;

Longspurs and **Snow Buntings** at Illinois Beach and other beaches and particularly in agricultural fields where manure has been spread.

February

This is a good month to look for owls and gulls. Continue to look for the same species as in January, but also look for the following:

Canvasbacks, Redheads, and **Common Mergansers** at Illinois Beach and inland glacial lakes of Lake County (Ill.);

All three **scoters** (particularly White-winged Scoter) at Illinois Beach, Waukegan Harbor, Tower Road, and Gillson Park;

Horned Larks in open fields;

Early **Red-winged Blackbirds** and **Grackles** in wetland areas.

March

Cranes, herons, and **ducks** return to the Chicago area. When the ice leaves, try any of the lakes in Lake County (Ill.), Moraine Hills, Chicago Botanic Garden, McKee Marsh, Fermilab, Palos preserves, Willow Slough, Jasper-Pulaski, Kankakee Fish and Wildlife, Kingsbury Fish and Wildlife, LaPorte, Valparaiso Lakes, and the Illinois River. In March, look for the following:

Herons as they return to Bakers Lake, Indiana Dunes, Lake Calumet, and Lake Renwick;

Greater White-fronted Geese at Fermilab, Kemper Lakes, Kingsbury, and Jasper-Pulaski and wherever there are large flocks of Canada Geese;

Tundra Swans on inland Lake County (Ill.) lakes;

Northern Pintails, the earliest dabbling duck to return;

Bald Eagles at Saganashkee Slough, Willow Slough, and Eagle Point at Pistakee Lake;

Raptors (including **Red-shouldered Hawks** and **Northern Harriers**) at Johnson Beach and Mt. Baldy;

Sandhill Cranes gathering at Jasper-Pulaski and flying north. They are sometimes seen flying over Johnson Beach, Buffalo Creek, Crabtree, and Moraine Hills;

American Woodcocks performing nocturnal displays at many sites, including Sauganash Prairie, Wolf Road Prairie, Spring Creek Valley, Cuba Marsh, Illinois Beach, and Ryerson;

Short-eared Owls migrating along Lake Michigan;

Early **passerine** migrants at lakefront parks and the Lake Calumet region;

Eastern Bluebirds at numerous sites, including Illinois Beach, Volo Bog, and Ryerson;

Longspurs on agricultural land with fresh manure. Fields near Willow Slough are especially good for Smith's Longspurs;

Rusty Blackbirds at Jasper-Pulaski and Third Lake and along the Prairie Path and other marshy areas.

April

Waterfowl, shorebirds, and other migrants continue to move into the region. Look for the following:

Common Loons at Moraine Hills, Pistakee Lake, Fox Lake, Long Lake, Flint Lake, and Deep Lake;

Double-crested Cormorants at Bakers Lake, Riverdale Quarry, and Lake Renwick;

Black-crowned Night-Herons returning to Indian Ridge Marsh, Big Marsh, and Lake Renwick;

Turkey Vultures at Saganashkee Slough, Red-Wing Slough, and refuges in the Kankakee River basin;

Ducks on inland lakes and sloughs, refuges in the Kankakee River basin, Baker's Lake, McGinnis Slough, small glacial lakes, Red-Wing Slough, Blackwell Forest Preserve, Prairie Spring Park, and Nicholson Pond;

Ospreys at Saganashkee Slough, Willow Slough, Moraine Hills, other Lake County (Ill.) lakes;

Broad-winged Hawks among the migrating raptors that can be seen at Johnson's Beach, Elsen's Hill, and James Park;

Virginia Rails and **Soras** appearing in late April in lakefront parks and in marshes throughout the area;

Sandhill Cranes still at Jasper-Pulaski, Glacial Park, and other nesting areas in Walworth County (Wis.), McHenry County and Lake County (Ill.);

American Golden-Plovers in the Enos, Ind., area;

Greater and **Lesser Yellowlegs** and **Pectoral Sandpipers** on mudflats and inland flooded fields;

Willets along Lake Michigan beaches from late April into early May;

Common Snipe in wet, grassy areas and roadside ditches at Moraine Hills, Spring Creek Valley, Glacial Park, Songbird Slough, Spring Lake Nature Preserve, Cuba Marsh, Illinois Beach, Chain O'Lakes, and Goose Lake Prairie;

Bonaparte's Gulls at Montrose Harbor, Skokie Lagoons, Waukegan Harbor, and inland lakes;

Common and **Forster's Terns** on the lakefront;

Great Horned Owl nestlings wherever they nest;

Purple Martins at Crabtree and at martin houses;

Tree Swallows along Lake Michigan and over inland lakes;

Winter Wrens in ravines and riparian woods;

American Pipits in late April at mudflats in open fields, Campbell Airport, and Illinois Beach;

Early **vireos** and **warblers** at Indiana Dunes, Dunes Heron Rookery, Wooded Isle, Magic Hedge, Tinley Creek Woods, Skokie Lagoons, and Chicago Botanic Garden;

Louisiana Waterthrushes at Lincoln Park Bird Sanctuary, McClaughry Springs, Ryerson, and Ft. Sheridan ravines;

Eastern Towhees at Illinois Beach and lakefront parks;

Sparrows (Vesper, Savannah, Le Conte's (rare), Song, and Swamp) at the Magic Hedge, Wooded Isle, McCormick Place, McKee Marsh, and Pratt's Wayne Woods.

May

May is the month birders wait for all year long. It is the height of migration for many birds including shorebirds, flycatchers, vireos, thrushes, warblers, tanagers, and grassland species. Birds are generally seen in the southern areas first, and the weather has a great influence in when species appear. Southerly winds help migrating birds, and an evening with southerly winds and clear skies is likely to encourage the movement of passerines. Listen for call notes of birds at night, and look for birds that may have landed early in the day. Birds that stray or are blown out over Lake Michigan at night come back to land in the morning. These "landfalls" can be spectacular. Strong winds from the southwest sometimes blow western species into the Chicago area. The Rare Bird Alert lists unusual species, but rarities seldom stay long in May. Many bird clubs and nature centers offer field trips in May, and various Spring Counts of birds take place in May. Volunteers are needed to take part, and a call to your local birding organization will put you in touch with the compiler for your county. Any location can be good for migrant passerines and warblers, but known "hot spots" include Illinois Beach, Chicago Botanic Garden, Wooded Isle, Magic Hedge, Morton Arboretum, Ryders Woods, Skokie Lagoons, Olive Park, Deer Grove, Palos area, Elsen's Hill, Pratt's Wayne Woods, Glacial Park, Moraine Hills, McKee

Marsh, Hidden Lake, Spring Creek Valley, Ryerson Woods, Lincoln Park Bird Sanctuary, Migrant Trap, Indiana Dunes, Michigan City, Warren Dunes, Forest Lawn, Warbler Walkway, and Chain O'Lakes. In May look for:

Double-crested Cormorants at Baker's Lake, Lake Renwick, and Riverdale Quarry;

Herons active at nesting colonies at Lake Renwick, Big Marsh, Indian Ridge, Baker's Lake, Dunes Heron Rookery, and Riverdale Quarry;

Waterfowl lingering on inland lakes and in the Lake Calumet area;

Northern Bobwhites at Beaver Lake, Plum Creek Forest Preserve, Willow Slough, and Newton County fields;

Shorebird migration reaching its peak in mid-May at Lake Michigan beaches, Waukegan Beach, Illinois Beach, McKee Marsh, McGinnis Slough, and mudflats in agricultural fields;

Forster's Terns at Grass Lake and Tichigan Marsh;

Black Terns over McGinnis Slough, Palos West Slough, Moraine Hills, East Loon Lake, Red-Wing Slough, and along the Fox River;

Alder Flycatchers at Grant Woods, Volo Bog, Cowles Bog, Spring Lake, Greene Valley, Beverly Shores, Tinley Creek Woods, Bachelors Grove, and Cap Sauers;

Willow Flycatchers at Volo Bog, Cuba Marsh, Tinley Creek Woods, Bachelors Grove, and Cap Sauers;

Loggerhead Shrikes at Midewin and Goose Lake Prairie;

Sedge Wrens at Chain O'Lakes, Oak Ridge, and Sunset Hill;

Veeries at Volo Bog and Chain O'Lakes;

Northern Mockingbirds at Beaver Lake, Midewin, Momence Sod Farms, and Des Plaines Conservation Area;

Dickcissels at Buffalo Creek, Beaver Lake, and Sunset Hill;

Bobolinks at Sunset Hill, Orland, Glacial Park, Spring Creek Valley, Buffalo Creek, Cuba Marsh, and McDonald Woods.

June

Look for nesting species this month. The breeding bird census takes place in June, and a call to your local birding organization will put you in touch with someone who needs your help. This month, look for:

Late **migrants** continuing to appear along the lakefront and at Greene Valley in early June;

Least Bitterns at Des Plaines Wetlands, North Dunes of Illinois Beach, Bong State Recreation Area, McGinnis Slough, Powderhorn Marsh, and Long Lake (Ind.);

Herons nesting in colonies at Big Marsh, Indian Ridge Marsh, Bakers Lake, Cedar Lake, Fox River, Lake Renwick, and Dunes Heron Rookery;

Shorebirds along Lake Michigan's beaches, Montrose Harbor, Waukegan Beach, Big Marsh, Deadstick Pond, McGinnis Slough, and O'Hare Wetland;

Upland Sandpipers at Midewin and Bong State Recreation Area;

Ring-billed Gulls nesting in colonies at Waukegan Harbor, Dresden Ponds, and Lake Calumet;

Chuck-wills-widows at Willow Slough and Mt. Baldy;

Whip-poor-wills at North Dunes of Illinois Beach, Camp Logan, Little Red Schoolhouse, Indiana Dunes, Jasper-Pulaski, and Warren Dunes;

Loggerhead Shrikes at Midewin, Goose Lake Prairie, Heidecke Lake, and Springbrook Prairie;

Bell's Vireos at Des Plaines Conservation Area, Fermilab, LaSalle Fish and Game, Braidwood Dunes, Springbrook Prairie, Midewin, Willow Slough, and Pembroke Township;

Sedge Wrens at Chain O'Lakes, Fermilab, Orland Hills, Springbrook Prairie, Glacial Park, Spring Bluff, Sunset Hill, Goose Lake Prairie, and Oak Ridge County Park;

Black-throated Green Warblers at Grand Mere;

Yellow-throated Warblers at Dunes Heron Rookery, Forest Lawn, and Kankakee Fish and Wildlife;

Prairie Warblers at Indiana Dunes;

Cerulean Warblers at Indiana Dunes, Kankakee River State Park, Warren Woods, Des Plaines Conservation Area, Tyler Creek, and Oak Point;

Prothonotary Warblers at Indiana Dunes State Park, Sarrett Nature Center, LaSalle Fish and Game, Kankakee Fish and Wildlife, and McHenry Dam;

Worm-eating Warblers at Grand Mere;

Louisiana Waterthrushes at McClaughry Springs, Indiana Dunes, and Warren Woods;

Kentucky Warblers at Pilcher Park;

Hooded Warblers at Indiana Dunes and Warren Dunes;

Yellow-breasted Chats at Cowles Bog, Cherry Hill, Thornton-Lansing, Chain O'Lakes, Cuba Marsh, Des Plaines Conservation Area, Fermilab, and along the Kankakee River;

Summer Tanagers at Willow Slough, Newton County, Des Plaines Conservation Area, and Braidwood Dunes;

Vesper Sparrows at Buffalo Creek, Campbell Airport, Des Plaines Conservation Area, Illinois Beach State Park, and Newton County;

Lark Sparrows at Braidwood Fish and Wildlife, Pembroke Township, Willow

Slough, Des Plaines Conservation Area, and Newton County;

Grasshopper Sparrows at Goose Lake, Springbrook Prairie, Fermilab, Newton County, Orland Hills, Glacial Park, Herrick Lake, Willow Slough, Illinois Beach, and Cuba Marsh;

Henslow's Sparrows at Goose Lake Prairie, Herrick Lake, Willow Slough, Orland Hills, Bong State Recreation Area, Springbrook Prairie, Braidwood Fish and Wildlife, and Glacial Park;

Blue Grosbeaks at Pembroke Township, Willow Slough, Midewin, Braidwood Dunes, and Beaver Lake;

Dickcissels at Springbrook Prairie, Orland Hills, Glacial Park, Braidwood Fish and Wildlife, and Fermilab;

Bobolinks at McDonald Woods, Orland Hills, Bong State Recreation Area, Midewin, Glacial Park, Fermilab, Herrick Lake, Hampshire Forest Preserve, and Springbrook;

Yellow-headed Blackbirds at Moraine Hills, Almond Marsh, Wadsworth Wetlands, and Delavan Inlet.

Orchard Orioles at Half Day Preserve, Saganashkee Slough, Pembroke Township, Braidwood Fish and Wildlife, LaSalle Fish and Game, Willow Slough, Beaver Lake, Newton County fields, and Glacial Park.

July

Shorebirds returning south are seen in many places, including McGinnis Slough, Deadstick Pond, Big Marsh, Hegewisch Marsh, McKee Marsh, Fermilab, Hidden Lake, Glacial Park, Calumet sewage ponds of the Metropolitan Sanitary District, O'Hare wetlands, Volo Bog, and such Lake Michigan beaches as Illinois Beach, Waukegan Beach, and Montrose Beach. Water levels must be low enough to expose mudflats in the inland locations.

Upland Sandpipers may be seen late in the month at sod farms at Momence, Wind Lake, DuPage County, and Shamrock Turf Farms.

August

Continue to look for migrant **shorebirds** in the same locations as June and July along with the following:

Pied-billed Grebes and **Blue-winged Teal** at McGinnis Slough;

Herons and **egrets** dispersing after breeding and perhaps being seen in the Des Plaines Wetlands and Lake Calumet area marshes;

Buff-breasted Sandpipers and **American Golden-Plovers** at Momence Sod Farms, Wind Lake Sod Farms, Shamrock Turf Nursery, Naperville Polo Fields, and Schneider Sod Farms;

Caspian Terns along Lake Michigan and inland lakes;

Nighthawks migrating south in large flocks;

Migrant landbirds late in the month at Illinois Beach, Waukegan Beach, Chicago Botanic Gardens, Skokie Lagoons, Magic Hedge, Lincoln Park Bird Sanctuary, Wooded Isle, Elsen's Hill, Cliffside Park, Spring Creek Valley, Hammond Lakefront Park, and Indiana Dunes.

September

This month is the peak for the fall **passerine** migration. Good spots to look include the Chicago Botanic Garden, Rosehill Cemetery, North Park Village, Chicago Lakefront, Skokie Lagoons, Elsen's Hill, Magic Hedge, Tinley Creek, Bachelors Grove, Spring Creek Valley, McDowell Grove, Cliffside Park, Hammond Lakefront Park, Indiana Dunes, Colonial Park, and Hidden Lake. One may also find:

Herons and **egrets** at McGinnis Slough and Des Plaines wetlands;

Jaegers (best time for Long-tailed) at Miller Beach;

Shorebirds at McGinnis Slough, Lake Calumet areas, and lakefront beaches;

Buff-breasted Sandpipers until mid-September at the same locations as in August;

Ospreys at Illinois Beach and over inland lakes;

Accipiters along the lake, Spring Creek Valley, and Mt. Hoy;

Broad-winged Hawks away from the lakefront at such places as Elsen's Hill, Mt. Hoy, and along the Fox River;

Falcons, including **Merlins** and **Peregrines,** at Illinois Beach and other lakefront locations. The peak of the Peregrine flights is the last two weeks of September;

Nelson's Sharp-tailed Sparrows in wet, grassy areas along the Chicago lakefront, Illinois Beach, Chicago Botanic Garden's Marsh Island, and Whiting Park.

October

Many rarities are seen in October and November. Call the rare bird alert to learn about them. This month is the peak of migration for **waterfowl, cranes,** and **raptors.** Look for the following:

Red-throated Loons off Beverly Shores;

Waterfowl at McGinnis Slough, Palos West Slough, Will-Cook pond, Long John Slough, Tampier Slough, Baker's Lake, Long Lake (Ind.), Paw Paw Lake, and such Lake Michigan locations as Miller Beach, Gillson Park, St Joseph, New Buffalo, and Michigan City;

Scoters in Lake Michigan after mid-October;

Migrating **raptors** along the lakefront, Illinois Beach, Chain O'Lakes, Cliffside Park, Fermilab, Gillson, Mt. Hoy, and Johnson Beach throughout the month. Species include **Northern Harrier, Merlin,** and **Peregrine Falcon**.

Bald Eagles at McGinnis Slough, Saganashkee Slough, Willow Slough, and Jasper-Pulaski;

Sandhill Cranes over the Chicago area and at Jasper-Pulaski;

Red Phalaropes in protected harbors in Lake Michigan in late October–early November;

Jaegers over Lake Michigan at Miller Beach (best), Gillson Park, Northwestern University, and Waukegan Harbor;

Franklin's Gulls along the lake and inland lakes;

Little Gulls with Bonaparte's Gulls on lakefront harbors;

Late **migrant passerines** in lakefront parks and inland wooded areas;

Eastern **Bluebirds** in small flocks at the Skokie Lagoons and Fermilab;

American Pipits at Illinois Beach and Montrose Beach;

Sparrows in grassy and weedy areas along Lake Michigan and at Cherry Hill, Fermilab, and McKee Marsh;

Rusty Blackbirds in wet woods and Jasper-Pulaski.

November

This is a good month for rarities. Call the rare bird alerts to learn what has been seen. Look for the following:

Common Loons at Lake Michigan, Navy Pier, Montrose Harbor, and Fox Lake;

Tundra Swans, Oldsquaw, and **Common Goldeneyes** at Lake Michigan;

Black Scoters and **Surf Scoters** through early December in flocks of scaup on Lake Michigan;

Mergansers at McGinnis Slough and Will-Cook ponds;

Late migrating **raptors** at Fermilab, Illinois Beach, McGinnis Slough, and Willow Slough;

Sandhill Cranes reaching peak numbers at Jasper-Pulaski;

Purple Sandpipers (very rare) at Waukegan Harbor, Navy Pier, Shedd Aquarium breakwater, Michigan City Harbor, and New Buffalo;

Jaegers and **Black-legged Kittiwakes** over Lake Michigan at Gillson Park, Michigan City Landfill, and Miller Beach;

Wintering gulls at O'Brien Lock and Dam, Lake Calumet, Michigan City, and Miller Beach;

Bonaparte's Gulls early in the month at Saganashkee Slough, Little Calumet River, and Montrose Harbor;

Little Gulls (rare) at Illinois Beach, Gillson Park, Waukegan, New Buffalo, Michigan City, St. Joseph, and other lakefront locations;

Short-eared Owls at Spring Creek Valley and Pratt's Wayne Woods;

Northern Shrikes in open areas;

Horned Larks, Lapland Longspurs, and **Snow Buntings** at North Point Marina, Montrose Beach, and rural areas;

Crossbills and other **winter finches** at Morton Arboretum, Lyons Woods, and Illinois Beach. White-winged Crossbills prefer spruces and hemlocks; Red Crossbills prefer pines.

December

This month, Christmas Bird Counts are held throughout North and Central America. Volunteers are needed to participate in the field as well as to count the birds at their feeders. Local bird clubs can provide information about the dates of the count in your area and about whom to contact. After the counts, the Rare Bird Alert lists rare birds found on the counts. During the month, look for the following:

Mergansers and other waterfowl where the water remains open in Lake Michigan, on other large inland lakes such as Lake Geneva, and on rivers below dams;

Raptors in open areas such as Bong State Recreation Area and Fermilab;

Bald Eagles at McGinnis Slough, Heidecke Lake, and Willow Slough;

Rough-legged Hawks at Illinois Beach, Indiana Dunes, and open areas in Kane and DeKalb Counties;

Wintering gulls in lakefront harbors and at O'Brien Lock and Dam, Lake Calumet Area, Michigan City, and St. Joseph and at landfills in the Calumet area and south of Michigan City;

Wintering owls at Bong State Recreation Area, O'Hare Airport, Chain O'Lakes, Morton Arboretum, and Vollmer;

Snowy Owls at Indiana Dunes, Meigs Field, and Navy Pier;

Snow Buntings at Indiana Dunes and on the beaches at Montrose Harbor and Illinois Beach;

Winter finches at Lyons Woods, Illinois Beach, Chicago Botanic Garden, Crabtree, and Morton Arboretum.

Birding Locations

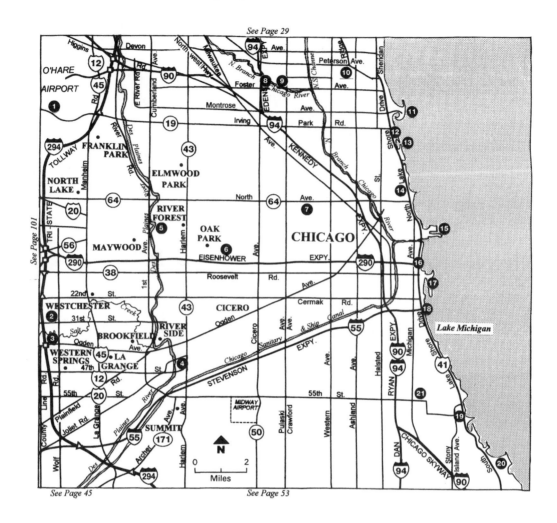

See Page 29

See Page 101

See Page 45

See Page 53

Map M   **Chicago and Central Cook County**

1. O'Hare Post Office Ponds *p. 23*

2. Wolf Road Prairie Nature Preserve *p. 25*

3. Bemis Woods Forest Preserve *p. 25*

4. Ottowa Trail Woods Forest Preserve *p. 24*

5. Thatcher Woods Forest Preserve *p. 24*

6. Columbus Park *p. 27*

7. Humboldt Park *p. 27*

8. LaBagh Woods Forest Preserve *p. 26*

9. Sauganash Prairie *p. 26*

10. Rosehill Cemetery *p. 27*

11. Magic Hedge and Montrose Harbor *p. 18*

12. Lincoln Park Bird Sanctuary *p. 18*

13. Belmont Harbor *p. 19*

14. Lincoln Park Zoo *p. 20*

15. Navy Pier *p. 20*

16. Grant Park *p. 20*

17. Meigs Field *p. 21*

18. McCormick Place *p. 21*

19. Jackson Park *p. 21*

20. Rainbow Beach *p. 23*

21. Washington Park *p. 28*

Chicago and Central Cook County

Chicago Lakefront Parks

Chicago's beautiful lakefront has become the city's defining feature, and nearly all of it is readily accessible to the public. Most of the 20-mile shoreline is comprised of fill (Chicago has steadily oozed east over the decades), and none of it is natural. But the miles of parkland, beaches, rocks, breakwaters, lagoons, harbors, and the lake itself provide some of the best birding in the Chicago region. Virtually every bird that regularly occurs in this region has at one time or another been seen on the Chicago lakefront, and more rarities have been found here than any other local area. The almost endless list of rarities includes Pacific Loon, Reddish Egret, Mew Gull, Groove-billed Ani, Swainson's Warbler, and Cassin's Sparrow.

W S S F

In winter, the land bird potential is limited, but Lake Michigan attracts occasional Harlequin Ducks (particularly inshore near piers and breakwaters), White-winged Scoters (which begin to appear in numbers as early as late February), dark-winged scoters, Oldsquaws, Common Goldeneyes, and Common and Red-breasted Mergansers. King Eiders (almost always females or young males) are possible. Gulls are plentiful, with Ring-billed and Herring congregating on harbor ice. These gatherings usually contain other species as well, including Thayer's, Iceland, Lesser Black-backed, Glaucous, and Great Black-backed Gulls. But if there is one single bird that is associated with the winter lakefront it is surely the Snowy Owl, for there is no more reliable place to find it in the area. In flight years, these owls are difficult to miss, but even in off-years, one or two usually spend all or part of the season.

More birders comb the lakefront parks during spring and fall than any other Chicago area sites. Their efforts are usually rewarded, and when west winds concentrate migrating land birds along the lake, the birding can be spectacular. The tiny Montrose area has yielded over 100 species in several hours, and a single observer once recorded 31 species of warblers one day in May at Jackson Park. Most passerines migrate from March though the beginning of June and again from August through November.

Because hedges and lawns are easier to bird than woods and marshes, many of the more secretive birds are best seen in lakefront parks. One spring morning at Montrose, birders encountered an American Bittern standing erect and still. Such a pose would have made it invisible in cattails, but the poor bird seemed unaware that it was standing in grass three inches high. Most birders who have seen Yellow and/or Black Rails in the Chicago area have done so on the lakefront. These parks are also among the best locations to see the skulking sparrows

(Henslow's, Le Conte's, and Nelson's Sharp-tailed) and the secretive Mourning and Connecticut Warblers.

Peregrine Falcons and Merlins tarry along the lakefront during migration to pursue feathered prey, and Purple Martins gather by the hundreds in August prior to embarking on their southward flights. A wide variety of sandpipers use the park beaches. Sanderlings are frequent, but American Avocets, Whimbrels, Buff-breasted Sandpipers, and other species also appear regularly. Another group of birds, more exclusively aquatic, migrate offshore in early spring and late fall. These include loons, grebes, Double-crested Cormorants, jaegers, and terns. Views of these species are often best on days with east winds. Jackson Park is the only one of the parks big enough to hold a variety of birds in the summer months, but early and late migrants, as well as sandpipers, can be found in the summer along the lakefront.

Hordes of people converge on these open spaces to partake of every imaginable activity. Thus to avoid dog walkers, cyclists, rollerbladers, and other park users, it is advisable to arrive as early in the morning as possible, particularly during the warmer months. Some places, like the heavily birded Magic Hedge and Lincoln Park Bird Sanctuary, are probably safe to bird alone, but solo birding at other sites should be avoided, particularly for those unfamiliar with the areas. The following sites are owned by the Chicago Park District unless otherwise indicated:

The Magic Hedge and Montrose Harbor (M11)   The 150-yard row of bushes and trees known as the Magic Hedge is one of the best known and best birded sites in the Chicago area. It is located on Montrose Point, a promontory in Lincoln Park that extends well into Lake Michigan. The Chicago Park District recognizes the importance of the Magic Hedge to migrant birds and has placed a sign at the hedge to mark this special area.

Directions—Go east of Lake Shore Drive on Montrose Avenue (4400N) and turn south (right) at the bait shop. The small rise with trees on the left is the Magic Hedge. Park along the road before it circles around Montrose Harbor on the right.

Trails—Walk across the drive to the east side and up to the bushes. You will probably see other birders, too. Walk the perimeter of the hedge slowly for the best birding, then walk east to Montrose Beach and the jetty. On-shore winds from the east seem to push migrant passerines to inland locations, but east winds bring jaegers and other water birds close to the shore in season. A beach house north of the hedge provides shelter from the winds.

Montrose Harbor is a small, protected marina on Lake Michigan that is attractive to Snowy Owls and numerous water birds. Continue south of the Magic Hedge, circle the harbor, and park near the turn-around at the end of the drive.

Trails—Bird from your car or walk along the edges of the harbor or along the lakefront.

Lincoln Park Bird Sanctuary (M12)   The sanctuary is a dense four-acre woods with pond and cattail marsh enclosed by a tall, cyclone fence that makes it the most sheltered spot in Lincoln Park. But, like everyone else, birders must be content

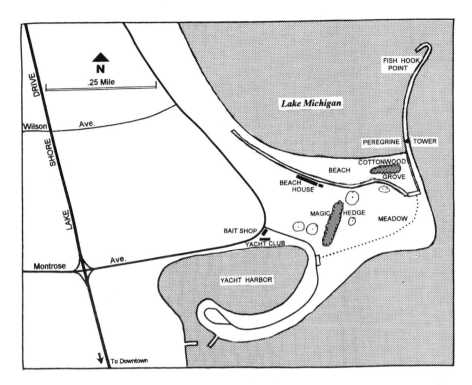

Magic Hedge and Montrose Harbor

with walking around and peering through the fence. (While you are peering, look for coyotes, which in recent years have moved into the neighborhood.) Winter birding has improved since birders began scattering seed around the fence. Taking advantage of the largesse, a Brewer's Sparrow survived the winter here several years ago. The archery range south and east of the sanctuary is also worth covering.

Directions—Exit Lake Shore Drive at Irving Park Avenue, turn south and proceed for .5 mile. Park near the totem pole. The sanctuary is behind the totem pole.

Trails—Walk around the perimeter fence and look in and up.

Belmont Harbor (M13)   This site consists of a sheltered harbor on Lake Michigan in Lincoln Park south of the Lincoln Park Bird Sanctuary. Gulls are present all year, and diving ducks remain as long as the water stays open. The peninsula between the harbor and the lake is often full of sparrows.

Directions—Exit from Lake Shore Drive onto eastbound Irving Park Road and turn south to the harbor. A number of parking lots are available. By parking at the north end of the harbor, you can also visit the adjacent Lincoln Park Bird Sanctuary. By continuing down the strip between the harbor and the lake, you can park looking out to the lake or into the harbor. The end of this peninsula is sometimes referred to as Gill Point or the Peninsula. Parking is also available at the south end of the harbor near the boat clubhouse off of Belmont Avenue.

Lincoln Park Zoo (M14)   This free zoo is one of the most popular attractions in the entire state. Its wooded grounds and duck ponds attract a wide variety of migrant land birds and waterfowl. Some of the latter spend the winter, but only zoo personnel know which are wild and which are part of the collection. Fortunately, they cheerfully share the information.

Directions—Exit Lake Shore Drive at Fullerton westbound. Park on Stockton Drive on the west side of the zoo or in the costly lot on Simonds Drive on the east side of the zoo south of Fullerton. Parking can be difficult in summer and after 8:00 A.M.

Trails—Walk the trails in the zoo and check the lagoons north and south of the zoo. North Pond is north of Fullerton, and South Pond is near the Farm in the Zoo.

Telephone—(312)742-2000

Navy Pier (M15)   One of Chicago's major tourist attractions, Navy Pier extends into Lake Michigan for a mile. It is thus an excellent vantage point from which to observe migrating and/or wintering loons, grebes, Oldsquaw, and other waterfowl. The old pier pilings south of the pier and the jetties to the east occasionally have Snowy Owls in winter. Nearby, **Olive Park** can be outstanding in migration. The hawthorn trees near the wall of the filtration plant often attract thrushes and sparrows in migration and in winter.

Directions—Drive or walk east of Lake Shore Drive on Grand Avenue to the pier. Olive Park is north of the pier between Lake Shore Drive and the Chicago Water Filtration Plant. Expensive parking is available at the end of Navy Pier. Try parking west of the drive.

Trails—Walk out to the end of the pier and scan the jetties and pilings south of the pier. For Olive Park, walk north to the wooded area and walk north along the path looking east into the fenced area. The open, grassy area and bushes to the north can also harbor migrants.

Owned By—City of Chicago

Grant Park (M16)   Sandwiched between Chicago's Loop and Lake Michigan, Grant Park is a latecomer to the list of great lakefront birding sites. Ornithologists from the Field Museum and others who work nearby bird the area on their lunch breaks and have found it to be a wonderful location. Stately elm and crabapple trees, a prairie garden, and a rose garden enhance the park's large expanses of grass. Thrushes, warblers, sparrows, and other migrants pass through the park in large numbers in spring and fall. The Grant Park area seems best during and immediately after major flights; unlike some of the other lakefront parks, it doesn't hold birds for extended periods.

Directions—The park is located between Michigan Avenue and Burnham Harbor and bordered by Randolph Street on the north and the Field Museum of Natural History on the south. Parking is available on Columbus Drive which cuts through the middle of the park.

Trails—Favorite birding spots include the prairie garden north of Monroe Street, the rose garden north of Buckingham Fountain, the trees east of the Field Museum, and the trees and shrubs on the south side of McFetridge Drive south of the Field Museum.

Meigs Field (M17)   This lakefront airport is the most reliable place for Snowy Owls in the Chicago area. Even in non-invasion years, at least one can usually be found at the airport or on nearby breakwaters between November and March. Short-eared Owls, Horned Larks, and Snow Buntings can also be found on or near the airport runways. Gulls may be found on nearby 12th Street Beach, in the harbors north and east of the field, and on the breakwaters. Though rare, Purple Sandpipers have been seen on the breakwaters in November. Other shorebirds are often found on the beach in migration.

Directions—Go east of the Field Museum of Natural History and the Shedd Aquarium on Solidarity Drive toward the Adler Planetarium and turn south into the parking lot for Meigs Field. **Monroe Harbor** is north of the peninsula, the **12th Street Beach** is at the tip of the peninsula, and **Burnham Harbor** is between Meigs Field and Lake Shore Drive. For a good view of Meigs Field, enter the terminal and walk upstairs to the observation platform.

McCormick Place (M18)   A huge convention center situated on Lake Michigan at 22nd Street, McCormick Place is one of the few buildings to interrupt the open parkland along the lakefront. Its strategic location and reflective walls lure large numbers of birds and bats to their deaths every year. Field Museum personnel collect the corpses each morning for an ongoing study of migration and bird mortality. Look for survivors in the grassy areas near the rocks and in the bushes and trees south of the building. The 31st Street beach south of the center can be good in migration for Franklin's Gulls.

Directions—Parking is difficult here. Pay to park at the convention center or park at the lot for the 31st Street beach and walk north.

Trails—Check all the shrubby areas, especially the upper area near the McCormick Place building. The grassy strip with four areas of shrubs and small trees south of McCormick Place can be particularly good.

Owned By—City of Chicago

Jackson Park (M19)   Built as part of the 1893 Columbian Exposition, this jewel of a park encompasses harbors, lagoons, islands, mature trees, thickets, shrubs, a marshy area, and large expanses of lawn. Within the park, the most interesting birding is at the Paul Douglas Nature Sanctuary (also known as Wooded Island), just south of the Museum of Science and Industry. Bobolink Meadow on the east side of the lagoon across from the island provides good habitat for grassland species and the shrubs along the fence between the meadow and the driving range are also worth checking. Black-crowned Night-Herons roost on the small islands east of Wooded Island during the spring and summer, and Bank Swallows often nest near La Rabida Children's Hospital in the southern part of the park. Other good areas for birds in Jackson Park are the wooded berm located a few blocks south (near the intersection of Cornell and 65th Street), the 63rd Street Beach, and "Promontory Point" which extends into Lake Michigan at 55th Street. Migrants are abundant in all locations, and ducks are often seen

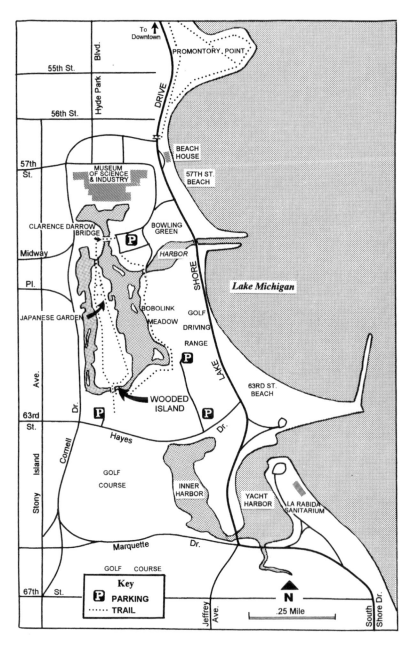

Jackson Park

from the beach and the point. Guided bird walks on Wooded Island are offered on Wednesday and Saturday mornings by a volunteer from the Chicago Audubon Society. (The walks are highly recommended, as they are led by an excellent birder who knows the park intimately.)

Directions—From southbound Lake Shore Drive, turn west into the park after passing 57th Street and the museum. Park in the lot near the "Clarence Darrow Bridge." Most bird walks begin at this bridge. If coming from the south, turn off

Lake Shore Drive onto 57th Street and drive around the east side of the Museum of Science and Industry to the lot on the south side. Other locations in Jackson Park south of Wooded Island can be reached from Hayes Drive (63rd Street). Turn west from Lake Shore Drive and park in the first lot for the 63rd Street beach, the second parking lot for the driving range and Bobolink Meadow, or the third lot for the "South Bridge" to Wooded Island. The wooded berm is south of Hayes on the east side of Cornell Drive. To reach it, drive south on Cornell from 57th Street or Hayes Drive until you see the wooded rise on the east side of the street. For Promontory Point at 55th Street, park near the Flamingo Hotel and walk east toward the lake via the underpass of Lake Shore Drive.

Trails—Walk across the Clarence Darrow Bridge and follow the path to the south where a second bridge, the "North Bridge" crosses to Wooded Isle. The shrubs and trees between the bridges often have interesting species. A paved path makes a loop around the island on the paved path. The light posts along the path are numbered in a clockwise direction beginning with Number 1 at the south end. Number 27 is at the north end near the North Bridge, and the highest number (Number 46) is back at the southern end of the island. Birders sometimes use these numbers to describe bird locations. The formal Japanese Garden is in the northeast part of the island. A popular walk is to enter the island at the north, circle through the island, exit the island over the bridge on the south end, head east, and then go back up north past the marsh and Bobolink Meadow between the lagoon and the driving range.

Rainbow Beach (M20)   This city park contains a sandy beach, trees, shrubs, a fenced grassy triangle, and a vegetable garden. Gulls frequent the beach and breakwaters, and shorebirds linger on the beach. In the southeast corner, the fenced vegetable garden is good in migration for Lapland Longspurs and various sparrows.

Directions—Follow South Shore Drive to Rainbow Park at 76th Street and park in the lot at the beach. There is another good beach south of the park and water filtration plant known as the 79th Street Beach.

Trails—Be sure to check the triangle grassy area at the south end of the beach.

Note—This location is best birded with others.

O'Hare Post Office Ponds (M1)

This site consists of an artificial complex of wetlands created for flood control south of the airport runways. As such, its long-term suitability for birds is in question. But for the time being, it is an excellent site for shorebirds from early April through mid-October, with the greatest variety in May, July, August, and September. Twenty-one species have been recorded, including American Golden-Plover, Black-bellied Plover, White-rumped Sandpiper, and Wilson's Phalarope. Horned Larks nested in 1993.

W S S F

Directions—Turn north from Irving Park Road onto Taft Road west of Mannheim Road. Turn right immediately and park in the parking area for the U.S. Post Office or bird from your car. Water levels fluctuate greatly from day to day. Because water drains here from the runways at O'Hare field, the smallest rainfall can change this area from a grassy marsh to a flooded pond within a very short time. The low area extends from west to east just north of Irving Park Road. To the west of Taft Road is a creek, pond, and area of mudflats; east of Taft, the pond extends into a marsh.

Owned By—City of Chicago

Des Plaines River

W S S F

The Cook County Forest Preserve District's extensive holdings along the Des Plaines River form a greenbelt that is sometimes referred to as Chicago's emerald necklace. In spring and fall, raptors and Sandhill Cranes migrate overhead while passerines are often abundant in the riparian woods. This is particularly true in spring, when the vegetation in these inland areas is up to two weeks more advanced than in lakefront sites. These woods of varying density also provide nesting habitat for Broad-winged Hawk, six species of woodpeckers, Great Crested Flycatcher, Eastern Bluebird, Scarlet Tanager, Rose-breasted Grosbeak, Indigo Bunting, and numerous other species. In winter, particularly if the river is open, these preserves often yield such birds as Brown Creeper, Golden-crowned Kinglet, and Yellow-rumped Warbler. Descriptions of two of the better preserves along the river follow:

Thatcher Woods (M5)   This wooded preserve and lagoon on the Des Plaines River is best in spring when northeastern winds disperse migrants inland. Warblers and vireos are plentiful, and Prothonotary Warblers are likely to be seen near the water in years when there is flooding. Tufted Titmice have nested here.

Directions—From the Eisenhower Expressway, exit at Harlem northbound and turn west on Lake Street. At Thatcher, turn north for two blocks and park in the lot for the **Trailside Museum**.

Trails—The lagoon behind the museum is attractive to birds in migration. Walk around the pond to the trail which goes south along the Des Plaines River to Madison Avenue. To the north, the trail continues north of Chicago Avenue into Thatcher Woods, skirts a parking lot and recreational field, then follows the riverbank. By continuing past North Avenue, you will come to Evans Field and Fullerton Woods.

Ottowa Trail Woods (M4)   This tract of mature floodplain forest and dense thickets on the east side of the Des Plaines River is the last remnant of the Chicago Portage, the low divide between the Great Lakes and the Mississippi River. Belted Kingfishers are present year-round. From mid-April through May, look for migrating warblers, thrushes, and other passerines. May is also a good time for both Black-billed and Yellow-billed Cuckoos in dense thickets and vines along the woodland edge. Five species of woodpeckers nest here, and American Kestrels nest near the oil refinery south of 46th Street. Other nesting species include Willow Flycatcher and Indigo Bunting in the thickets and brushy areas along the railroad tracks. The prudent visitor should be prepared not only for birds, but mud and mosquitoes as well.

Directions—Enter the preserve from Harlem at 44th Street or 46th Street. Park in one of the lots and bird the areas between the drive and the river. To see the west side of the river, turn south from Joliet Road at 43rd Street into Stony Ford Woods and park at the south end of the drive. Canoes can be launched from Stony Ford to follow the Chicago Portage Canoe Trail downstream to Columbia Woods or Isle a la Cache.

Trails—There is one trail which heads north; another trail circles close to the river and extends to the south. On the west side of the river, follow the trail south along the river, cross 47th Street, and continue walking south along the river. Or, after crossing 47th Street, walk along the railroad bridge to cross the river to the east side. (Use extreme caution if walking on the railroad tracks— trains have the right of way!) **Chicago Portage Woods**, a National Historic Site, is south of the preserve on the south side of the railroad tracks, and **Stony Ford Woods** is across the river.

Wolf Road Prairie Nature Preserve (M2)

W S S F

The black soil prairie protected here is one of the finest surviving remnants in the entire state. The sidewalks that crisscross the prairie are reminders of how close the area came to being destroyed before the Great Depression wiped out the developers in the nick of time. Besides prairie, these 80 acres are enriched by open oak woods and marsh. In spring, American Woodcock engage in their crepuscular nuptial displays, and Virginia Rails and Soras are found in the marsh. Breeding species have included Cooper's Hawk, Tree Swallow, Eastern Bluebird, Yellow-breasted Chat, and Indigo Bunting.

Directions—Park in one of the three paved areas on the north side of 31st Street west of Wolf Road. Each area holds five cars. If the parking areas are full, park in the shopping center on the northeast corner of 31st Street and Wolf Road.

Trails—Sidewalks traverse the area. The bur oak savanna is in the southern portion, the native prairie grassland is in the center, and the marsh is at the northern edge of the preserve.

Note—Traffic is heavy on 31st Street. Stay on the sidewalks to preserve the prairie plants.

Owned By—Forest Preserve District of Cook County and Illinois Department of Natural Resources

Bemis Woods (M3)

W S S F

Over half of this 400-acre woods has been designated as **Salt Creek Woods Nature Preserve**. Rewarding during migration, Bemis Woods has also attracted such nesters as both species of cuckoo, Blue-gray Gnatcatcher, Veery, and Yellow-breasted Chat.

Directions—For the areas north of Salt Creek, drive onto the entrance road on the west side of Wolf Road between 31st Street and Ogden Avenue. Park in the lot near Salt Creek. For the areas south of Salt Creek and the Salt Creek bicycle trail, enter from Ogden just east of I-294. Park in the lot near the toboggan slides for the woods and bicycle trail or continue further west for the marshy area.

Trails—By walking south through the open, grassy area ("practice slopes"), you will reach a marshy area near Ogden. A bridge crosses Salt Creek and traverses the woods. A bicycle trail crosses into the part of the preserve east of Wolf Road and continues as far as **Brookfield Zoo**.

Owned By—Forest Preserve District of Cook County

North Branch of the Chicago River

W S S F

The grasslands and woods that comprise the forest preserves along the North Branch of the Chicago River can be excellent for migrant land birds, especially warblers and sparrows. Prothonotary Warblers, Worm-eating Warblers, Louisiana Waterthrushes, Hooded Warblers, and Le Conte's Sparrows have all been found here. The following sites are owned by the Forest Preserve District of Cook County:

Sauganash Prairie (M9)   This preserve of woods and restored prairie is an excellent place to observe the courtship displays of both Common Snipe and American Woodcock. (A particularly good place is from the shorter trail in the northeastern section.) Broad-winged Hawks and Red-Headed Woodpeckers have nested west of the main trail.

   Directions—Go south from Peterson on Kostner to Bryn Mawr and turn west. Park near Kilbourn.

   Trails—The main trail heads south from Bryn Mawr and Kilbourn. It borders the prairie on the left and parallels the Chicago River before passing under railroad tracks. Once beyond the tracks, you are in LaBagh Woods. Another shorter trail one-half block east of the main trail leads into the restored section of the prairie.

LaBagh Woods (M8)   The wooded bluffs along the river are particularly good for warblers in spring.

   Directions—For the north side of the river, go south from Peterson on Kostner and park on Bryn Mawr near Kostner. For the south side of the river, go east of Cicero at the Forest Preserve sign just north of Foster Avenue and park in the middle lot near the river.

   Trails—A railroad embankment crosses the river and connects the two portions of LaBagh Woods, so it is possible to park on either side to bird the whole preserve. If time is limited, bird the north portion which is less heavily used by the public. Walk the dirt paths along the river on both banks and circle around the pond between the river and the south parking lot. Bird the wet prairie in the northeast portion, and bird on either side of the tracks.

Rosehill Cemetery (M10)

In addition to headstones and graves, Rosehill Cemetery contains 350 acres of grass, shrubs, trees, and pond in the midst of residential and commercial development. This island of greenery attracts many migrants, especially warblers in spring.

W S S F

Directions—Enter the cemetery from Ravenswood Avenue south of Peterson Avenue. The pond is in the northwest corner of the cemetery.

Telephone—(773)561-5940

Inland Parks

Although not as productive as the lakefront parks, a number of Chicago parks located inland have mature trees and wetland areas that attract birds in migration. A few even provide breeding habitat. These parks are not birded often, so it is advisable to visit them with companions.

W S S F

Humboldt Park (M7)   This park has a marsh, swimming pond, and lagoon. During spring migration, Pied-billed Grebes, ducks, Hooded Mergansers, American Coots, rails, Ruby-throated Hummingbirds, and warblers may show up here. Rails have been seen on the west side of the marsh south of the swimming pond. Nesters include Eastern Kingbird, Yellow Warbler, and Common Yellowthroat.

Directions—Enter the park from North Avenue east of Kedzie. Turn south on Humboldt Drive which traverses the park from north to south. The lagoon can be seen on both sides of the drive. Turn east on Grower Drive to walk around the lagoon and cattails.

Trails—Walk around the lagoon early in the morning during May or June to search for rails, coots, and grebes.

Telephone—(312)742-7549

Columbus Park (M6)   In addition to a golf course, the park has a cattail marsh, a lagoon, a waterfall, and ponds that attract waterfowl (as long as the water is open), rails, herons, and shorebirds (particularly in spring). Migrating warblers frequent the trees surrounding the rock pond and lagoon, and sparrows forage in the unmowed grass around the golf course fences. Nesting birds include Red-eyed Vireo, Warbling Vireo, and Chimney Swift. The swifts often drink, bathe, and feed on the wing at the north end of the lagoon.

Directions—Exit the Eisenhower Expressway at Central northbound. Enter the park on the west side of the road at Jackson Boulevard just south of Adams Boulevard. Turn south to the field house and golf shop. Park in a lot close to the golf shop or on the east side of the drive south of the field house.

Trails—Ask in the golf shop if you can bird hole #8, which has a cattail pond. Walk east toward the lagoon making sure to check the rock pond and waterfall south of the field house. Proceed south along the fence toward the pond on the golf course and beware of flying golf balls. Chicago policemen may be present: let them know you are birdwatching.

Telephone—(312)746-5046

Washington Park (M21)   This large park has a lagoon at the south end and Loredo Taft's sculpture "The Fountain of Time" at the southeast entrance. Warblers and other passerines often arrive here earlier in spring than at other Chicago locations. Black-crowned Night-Herons and Common Moorhens have summered in the park, and Pied-billed Grebes, Green Herons, American Coots, Monk Parakeets, Yellow Warblers, and Common Yellowthroats have nested.

Directions—Enter the park by turning west of Cottage Grove Avenue south of 59th Street. Drive across the south end of the park on Best Drive and view the lagoon from your car.

Trails—Bird from your car.

Telephone—(312)747-6823

Northern Cook County

Lake Michigan Shore

W S S F

Most of the Cook County lakefront north of Chicago is off limits to the public. The major exceptions are municipal parks and the campus of Northwestern University. While these sites are often excellent for land bird migrants, their principal attraction is the array of waterfowl that winter and/or migrate offshore. If the lake remains unfrozen, Greater Scaup, scoters, Oldsquaws, Common Goldeneyes, mergansers, and gulls linger throughout the winter at varying distances from the shoreline. Examining these flocks for rarities is a favorite winter pastime of hardy birders. The fall migration of water birds and raptors exceeds that of spring. When fierce northeasters bring the opening blows of winter, birders huddle for hours in their cars or other sheltered spots as they scan the lake for Red-throated Loons, scoters, eagles, falcons, jaegers, Sabine's Gulls, Black-legged Kittiwakes, Short-eared Owls, and a myriad of other possibilities. This type of birding may not be for everyone, but it can be very exciting. A few of the rarities that have been observed along this stretch of the lake include Ancient Murrelet, Ross's Gull, Burrowing Owl, Sage Thrasher, and Lark Bunting.

Northwestern University (K16)   Northwestern University's sprawling Evanston campus is rich with woody vegetation and includes beaches and a lagoon on a landfill that extends into the lake. Record numbers of Peregrine Falcons and Merlins have been seen in late September and early October, and the landfill's grassy expanse attracts shorebirds, Short-eared Owls, Horned Larks, American Pipits, Lapland Longspurs, Snow Buntings, and rare sparrows in both spring and fall. (Something to think about as you bird the landfill is that the sand beneath your feet used to be the best part of the Indiana Dunes; the university purchased sand from the steel companies that leveled the central dunes.)

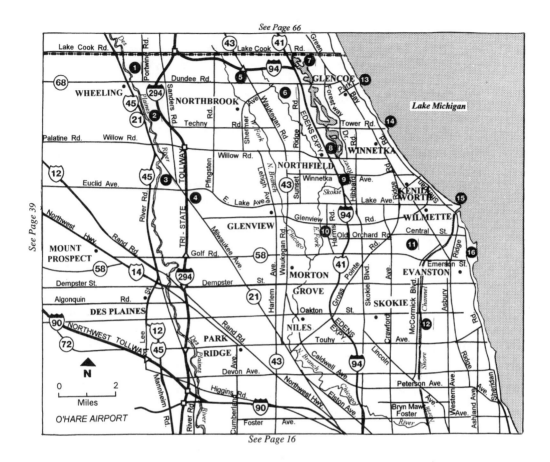

See Page 66

See Page 16

Map K   **Northeastern Cook County**

1. Portwine Road *p. 38*

2. Dam No. 1 Forest Preserve *p. 38*

3. River Trail Nature Center *p. 37*

4. The Grove *p. 36*

5. Somme Prairie *p. 32*

6. Chipilly Woods Forest Preserve *p. 32*

7. Chicago Botanic Garden *p. 35*

8. Skokie Lagoons Forest Preserve *p. 33*

9. Winnetka Avenue Bridge *p. 33*

10. Harms Woods Forest Preserve *p. 32*

11. Perkins Woods Forest Preserve *p. 31*

12. North Shore Channel *p. 31*

13. Park Avenue *p. 31*

14. Tower Road Park *p. 30*

15. Gillson Park *p. 30*

16. Northwestern University *p. 29*

Directions—For the north end of the landfill, go east of Sheridan Road on Lincoln Street and proceed east and south and out to the lake. Park along the road, but be aware that cars without stickers may be ticketed here during the week. Try to do your birding before 8:00 A.M. before the ticketing begins. To bird the Clark Street Beach and the south end of the landfill, turn east where Sheridan Road makes a right angle on the south side of the campus and park in the parking garage. Cars in the garage are not ticketed on weekends.

Trails—Walk the path that parallels the lakefront. The point of land at the north end is a good vantage point for the lake and birds migrating along the lake. The beach north of the landfill is fenced but can be scanned from the hill at the north end of the landfill. The evergreens east of the playing fields and the shrubs and trees around the lagoon offer excellent shelter for passerines.

Telephone—(847)491-5000

Note—The lake can also be viewed from the **Evanston Art Center** north of the campus.

Gillson Park (K15)   This village park is located where the Illinois shoreline juts farthest into the lake. Birds migrating offshore, therefore, are closer to land here than anywhere else. As such, this is a superb place from which to observe fall migration. Other birds, including large numbers of passerines, take advantage of the pier (a good place for Harlequin Ducks and Purple Sandpipers), sandy beach, harbor, tall trees, and wildflowers within the park. Ducks are often in the harbor all winter.

Directions—Enter the park from Lake Avenue or from Sheridan Road just north of the Baha'i Temple and the bridge over the North Shore Channel. Follow the drive north and park at the beach house. Parking is restricted during the summer months. To be safe, park on Michigan Avenue on the west side of the park.

Trails—The harbor is at the south end of the park where parking is available in non-boating season. Jetties extend out into the lake near the harbor entrance, and a sandy beach extends the length of the park. The open waters of Lake Michigan can be viewed from your car, or they may be scanned from the lee of the beach pavilion. Land birds often congregate in the area around the amphitheater.

Owned By—Village of Wilmette

Telephone—(847)251-2700

Tower Road Park (K14)   With its high bluff, public beach, and fishing pier extending into Lake Michigan, this park is a good place to see migrants moving along the shore in the fall and gulls and ducks in the winter.

Directions—Drive east of Sheridan Road to the parking lot at the terminus of Tower Road to see the beach and pier. For a higher vantage point, drive or walk east on the drive just south of Tower Road which ends on top of the coal chute for the power plant.

Owned By—Winnetka Park District

Telephone—(847)501-6000

Park Avenue (K13)This is another excellent overlook from which to observe water birds, especially goldeneys and scoters, in fall and winter.

Directions—From Sheridan Road, drive east on Park Avenue in Glencoe to the bluff overlooking the lake. Park in the lot on Park Avenue.

Owned By—Glencoe Park District

Telephone—(847)835-4114

North Shore Channel (K12)

A man-made canal bordered by deciduous and coniferous trees, fields, and ponds, it extends from Wilmette Harbor on the north to the North Branch of the Chicago River just south of Foster Avenue. Warm water discharged into the canal at Howard Street usually keeps the canal open south of this spot throughout the year. In winter, this section often yields waterfowl, gulls, Belted Kingfishers, and Black-crowned Night-Herons. Migrants can appear anywhere along the canal, but **Harbert Park**, **Thillens Woods**, and the section between Peterson and Argyle are particularly good for passerines. Cinnamon Teal and Black-throated Gray Warbler are two of the more unexpected species that have shown up here. **James Park**, a 45-acre city park on the east side of the North Shore Channel at Oakton, has man-made Mount Trashmore hill, a good vantage point for raptors in migration (March–May and September–November). In early November, Northern Shrike is a possibility in the area between Dempster and Devon. A surprising number of species have nested along the canal including Green Heron and Black-crowned Night-Heron, Red-shouldered Hawks, both cuckoos, Belted Kingfishers, Willow Flycatchers, Warbling Vireos, Brown Thrashers, Field Sparrows, and Eastern Meadowlarks.

Directions—The channel is bordered on the west by McCormick Boulevard north of Devon and by Kedzie south of Peterson. Places to observe include the bridge over the channel at Isabella Street or Harbert Park north of Main Street in Evanston, Thillens Woods north of Devon, or the bridges across the channel at Touhy, Bryn Mawr, Foster, or Argyle.

Owned By—Primarily the Metropolitan Water Reclamation District, but portions are leased to and managed by local municipalities

Perkins Woods (K11)

This site is a 7.5-acre Forest Preserve in the middle of Evanston, replete with vernal ponds and wildflowers. It is a tiny patch of virgin forest, the last remnant of what early settlers called "The Big Woods." Such animals as martens and fishers lived here 150 years ago. Perkin's Woods is popular with birders because it attracts large numbers of migrant land birds, including flycatchers, vireos, thrushes, and warblers. The tract is so small and accessible that birds are generally easy to find.

Directions—From Edens Expressway, turn off at Old Orchard Road and go east to Crawford Avenue. Turn south and proceed one block to Colfax. Drive east on

Colfax for nine blocks and park near the intersection of Colfax and Ewing Streets.

Trails—Paved paths crisscross this small site and meet in the middle.

Owned By—Forest Preserve District of Cook County.

North Branch of the Chicago River

W S S F

These riparian preserves protect woods, wetlands, re-created savanna, and a high quality prairie. Some feature superb displays of spring wildflowers, and all can be very productive for migrant land birds. The sites are owned by the Forest Preserve District of Cook County.

Harms Woods (K10)   These woods are particularly good in spring when carpets of wildflowers compete with birds for the attention of birders. Vernal ponds attract Gray-cheeked and Swainson's Thrushes, and damp brushy areas produce Connecticut, Mourning, and Canada Warblers. Pine, Prothonotary, and Kentucky are among the rarer warblers seen here on a regular basis. Probable nesting species include Veery in the woods on the west side of the river, Eastern Phoebe under the bridge, Blue-winged Warbler in the brushy wetland areas on the west side of the river, and Red-headed Woodpecker in dead tree trunks. Cooper's Hawks, Eastern Screech-Owls, and Great Horned Owls are year-round residents.

Directions—Four parking lots off of Harms Road offer access to trails on both sides of the river. The lot just south of Glenview Road for **Glenview Woods** is the closest lot to the footbridge over the river.

Trails—Stay on trails in Harms Woods to protect the vegetation. The bicycle trail along the east side of the river can be reached from any of the parking areas. To reach the trails on the west side of the river, walk across the bridge and turn south on the bridle trail which parallels the river. Approximately .5 mile south of the bridge, there is a fork in the trail; walk either trail as they form a loop approximately .5 mile further south. One trail follows the river; the other passes through brushy wetland and woodland.

Chipilly Woods Forest Preserve (K6)   The woods surrounding the river offer much the same birding opportunities as Harms Woods.

Directions—Three small parking areas exist. The first, on Lee Street approximately .5 mile south of Dundee Road, is the furthest to the north. The second is where Grant Street makes a right-angle turn .3 mile east of Lee Street, and the third area is on Voltz Road where the east-west road turns south.

Trails—A bridle trail parallels the river from Lee Road to the Voltz Road parking area. Trails lead to it from both of the other parking areas. The trail affords views into the floodplain where the best birding occurs.

Somme Prairie Complex (K5)   This site encompasses a high quality prairie that is dedicated as an Illinois Nature Preserve and a re-created savanna area known as Vestal Grove. Railroad tracks traverse the complex and divide the prairie from the savanna.

Directions—Drive west of Edens Expressway on Dundee Road and cross Waukegan Road. Park in the post office parking lot on the north side of Dundee Road for the Illinois Nature Preserve behind the post office. For Vestal Grove, return to the intersection of Dundee Road and Waukegan Road, turn north, and park in the unpaved lot on the west side of Waukegan Road. **Somme Woods** Forest Preserve, a dense stand of upland woods, is across the road on the east side of Waukegan Road. A trail circles the preserve and connects with **Chipilly Woods** to the south.

Skokie Lagoons Complex

W S S F

When the Wisconsin Glacier receded roughly 10,000 years ago, among the things it left behind was the Skokie Swamp, a large wet area that formed in the swale between two moraines. Fed and drained by the Skokie River, this inaccessible haven for wildlife survived until the 1930s, when crews from the Civilian Conservation Corps converted the rich wetlands into a chain of lagoons and islands. This area is now the Skokie Lagoons, and it offers superb birding throughout most of the year. A more recent addition is a ditch carrying treated wastewater that stays warm throughout the winter, running parallel to the lagoons near Edens Expressway and emptying into the Skokie River below the dam at Willow Road. We have divided the lagoons complex into three sections; the discussion proceeds from south to north. The entire complex is owned by the Forest Preserve District of Cook County.

Winnetka Avenue Bridge (K9)   The bridge on Winnetka Avenue marks the southern boundary of the Skokie Lagoons Complex. North of the bridge, mature deciduous trees, thickets, and seasonally wet fields flank both sides of the Skokie River. Blue-gray Gnatcatchers, Baltimore Orioles, and other species often appear here sooner in spring than at other locations in the lagoons. Later in spring, the woods can be excellent for many migrants including Philadelphia Vireos, and Connecticut, Mourning, Hooded, Wilson's, and Canada Warblers. The sedge meadow near the diagonal path has sheltered rails and bitterns in migration.

Directions—Park west of Hibbard Road on Winnetka Avenue near the bridge over the river.

Trails—Walk north on the bridle path on the east side of the river to Willow Road. Another diagonal path to the northeast along a small stream passes low, wet areas and ends at Hibbard Road. Once you reach Willow Road, a short hike to the west will take you to a paved bicycle path on the west side of the river. This path can be followed back to your car or taken across Willow Road to the Skokie Lagoons.

Skokie Lagoons (K8)   The best birding in the lagoons is in spring which arrives when the ice breaks up and waterfowl of many species dot the newly open water. Ring-necked Ducks, Buffleheads, all three mergansers, Ruddy Ducks, and gulls are often common. In early April, hundreds of Bonaparte's Gulls sometimes descend on the narrow lakelets, and later in the month, migrant land birds such as Yellow-bellied Sapsuckers, flycatchers, swallows, kinglets, thrushes, thrashers, tanagers, and sparrows arrive in large numbers. In May, the area is outstanding for cuckoos, vireos, and warblers. Many field trips are held here, and there are days every year when observers tally 25 or more warblers in a few hours. Yellow-crowned Night-Herons, Western Kingbirds, and Summer Tanagers show-up occasionally. Olive-sided and Yellow-bellied Flycatchers are seen each year in late May and again in early fall. Though less concentrated than in spring, throngs of migrants begin reappearing in mid-August and continue through November. Some birds take advantage of the microclimate near the warm water ditch by staying into winter. These have included Common Snipe, Winter Wren, Ruby-crowned Kinglet, Yellow-rumped Warbler, White-throated Sparrow, and on one occasion, an extraordinary Swainson's Thrush.

Directions—To access the southern areas, park in the lot on the north side of Willow Road just east of Edens Expressway. The forest preserve surrounding the lagoons in this area is known as **Erickson Woods**. Parking is also available along Forestway Drive on the east side of the lagoons and in lots on the south side of Tower Road. The closest parking to the ditch and northern end of the lagoons is west of Edens Expressway and south of Dundee Road.

Trails—From the Willow Road parking lot, walk east, cross the bicycle path, and enter the meadow. Walk north along the water's edge or return to walk north on the bicycle path. (Beware of the many bicyclists who travel the paths at high speeds!) The path continues to a parking lot at Tower Road and can be followed all the way up to Dundee Road on the west side of the lagoons (a 2.5-mile walk). A good circuit from Willow Road is to head north along the water's edge or bicycle path and turn back before reaching Tower Road. Look for the bridge over the warm water ditch on your left as you approach Edens Expressway. Take the bridle trail back south to the parking lot. To bird the island north of Tower Road, park on Forestway Drive just north of the picnic shelter .3 mile south of Dundee Road. Walk west along the north edge of the lagoon and follow the trail over the isthmus onto the island. The island extends south almost to Tower Road and can be quieter than the bicycle paths. North of Tower Road, the west side of the lagoons can be birded from the bicycle path. Look for the fenced marsh. Its grassy berms are attractive to birds, and the fence provides a convenient perch for insect-hunting birds. From Dundee Road southward, the warm water ditch along Edens Expressway stays open all winter.

Note—**Crow Island Park** on Willow Road, 2 blocks east of Hibbard Road, can be excellent in spring for woodland migrants.

Skokie Lagoons

Map labels

CHICAGO BOTANIC GARDEN
Hohlfelder Rd.
Dundee
N
.25 Mile
Skokie Blvd.
EDENS
Rd.
Sycamore
Lincoln
Ave.
Forestway
P
DIKE
LITTLE HOUSE OF GLENCOE
DAM
EXPRESSWAY

Key
..... TRAIL
P PARKING

DAM
Tower
P
Rd.
94
DAM
EDENS
Central
41
Forestway Dr.
Rd.
P
Willow Rd.
CROW ISLAND SCHOOL
Happ Rd.
EXPRESSWAY
Lagoon Dr.
Skokie River
Hibbard
Winnetka
Ave.
To Chicago

Chicago Botanic Garden (K7)   The garden consists of 300 acres of lagoons, oak woodlands, re-created prairies, and a wide variety of ornamental plantings. Weekend visitors and weekday school groups can make the place uncomfortably crowded, so it
is best to bird here early in the morning. A remarkable assortment of birds have occurred here: waterfowl, raptors, warblers, sparrows, and shorebirds can all be
counted on at the appropriate season. It is a good spot for rarities, among which
have been Scissor-tailed Flycatcher, Vermillion Flycatcher, and Townsend's Warbler.
Wintering ducks can be found in the open water that generally persists south of
the Sensory Garden. In invasion years, winter finches such as Red and White-
winged Crossbills, Common Redpolls, and Pine Siskins have been found in the
pines, firs, and birches, and at feeders. Migrant waterfowl are plentiful, with the

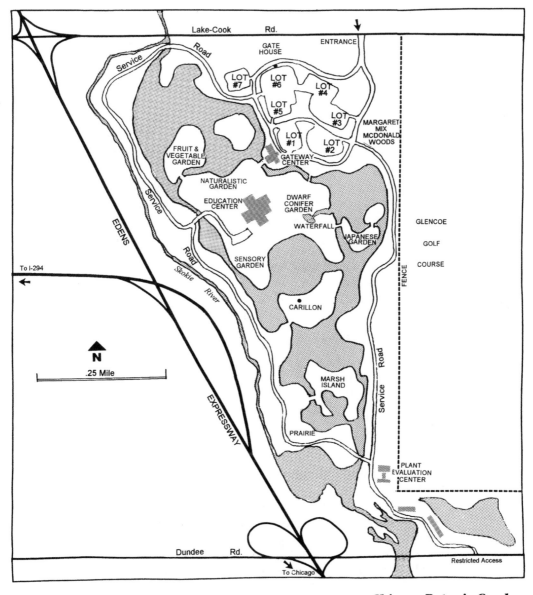

Chicago Botanic Garden

greatest variety present during the first half of April and the month of October. Beginning in early March, flocks of migrating passerines and raptors can be seen flying north in large numbers. The first three hours after daybreak in the prairie can be an excellent place from which to observe these movements. Sparrows and grassland birds are found in the prairie and along the west side of the garden, and rails have been seen along the boardwalk on Marsh Island. Warblers can be numerous along the northern perimeter, in Turnbull/McDonald Woods, in the Sensory Garden, and along the eastern fence line. Black-crowned Night-Herons, Willow Flycatchers, and Orchard Orioles are among the more interesting species that have nested on the property. Shorebirds are attracted to the exposed mud at the south end of Marsh Island, and Black-bellied Plover, Lesser Yellowlegs, and Dunlin have lingered into November. For the fall raptor migration, go to the Dwarf Conifer Garden and scan the skies to the north. A scope will be helpful.

Directions—Enter the Botanic Garden off Lake-Cook Road .5 mile east of Edens Expressway. If approaching from northbound Edens Expressway, be careful to stay on Rte. 41 (DO NOT enter the ramp for I-94 to Milwaukee). From the parking lots, Turnbull Woods (recently renamed the Margaret Mix McDonald Woods) is to the east, the Gateway Visitors' Center and demonstration gardens are to the southwest, and the prairie and Marsh Island are to the south. Nonmembers must pay a fee to park after 8:00 A.M. ($5.00 weekdays, $6.00 weekends in 1999). The entrance gate is open from sunrise to sunset, and maps are available there or in the Gateway Center.

Trails—A paved road encircles the gardens and offers good views of the lakes. This road is closed to private vehicles, but it may be walked in either direction. Guided tram tours follow the road and stop at a few areas, but they are not bird tours. Paths offer access to other areas of the garden. Favorite birding areas include Turnbull/Margaret Mix McDonald Woods, Marsh Island, the Prairie, the fence along the eastern boundary, the Sensory Garden, the waterfall, the Dwarf Conifer Garden, the Naturalistic Garden, and the feeders near the Education Center and the plant evaluation garden. The Skokie River is being restored to a more natural state and may be attractive to marsh and shorebirds in the future.

Telephone—(847)835-5440

Note—The **Turnbull/McDonald Woods Forest Preserve** is east of the Botanic Garden on the east side of Green Bay Road.

The Grove (K4)

W S S F

This is an 85-acre wooded preserve that was the family home of Robert Kennicott (1835-1866), northeastern Illinois's first naturalist. It is also the setting of Donald Culross Peattie's *A Prairie Grove* (1938), which is an extraordinary blend of natural and cultural history. (Peattie married a Kennicott.) An interpretive center with live animals is used by many school groups. Flycatchers, vireos, thrushes, and warblers can be expected in May and again in September. During late May, Connecticut and Mourning Warblers are usually present, as are Olive-sided Flycatchers that make hunting forays from the tops of dead trees near the parking lot. Yellow-billed Cuckoos have been seen on the far end of the long boardwalk by the fence. Nesting Wood Ducks, Blue-winged Teal, Red-shouldered

Hawks, Eastern Wood-Pewees, Great-Crested Flycatchers, and White-breasted Nuthatches make this an interesting summer destination.

Directions—From East Lake Avenue, drive south on Milwaukee Avenue, go under the I-94 underpass, and turn east into the drive marked with the large sign for The Grove. Pass the historic Kennicott house and park in the lot near the interpretive center. Because the site hosts many events, it is a good idea to call before visiting to avoid the crowds and parking fees imposed on certain weekends.

Trails—Trails and boardwalks make loops throughout the property. The trails are not marked, and it is easy to get lost. To be on the safe side, take the short loop around the pond outside of the interpretive center. Try the trail that goes behind the Indian camp. By following it, you will come to a long boardwalk over flooded woods and a trail that goes into a little-used part of the preserve. Here the trails are confusing, so turn around after crossing the boardwalk or carry a compass to help you find the way back.

Note—Brush removal begun in 1998 is likely to affect the birds seen here.

Owned By—Glenview Park District

Telephone—(847)299-6096

Des Plaines River Sites

The Cook County Forest Preserves north of Chicago along the Des Plaines River offer good land birding during spring and fall. Likely species include a wide variety of flycatchers, thrushes, vireos, and warblers. The mature woods, backwaters, and old fields also harbor a variety of typical nesting species. By far the most noteworthy animal here is the massasauga rattlesnake, a once common species that is barely hanging on. Due to its rarity and shyness, the snake is rarely encountered, but its very presence enhances the quality of these preserves. All sites are owned by the Forest Preserve District of Cook County.

W S S F

River Trail Nature Center (K3)   In addition to the many species that can be seen in migration along the river and its backwaters, the following are known to summer here: Great Blue Heron, Great Egret, Green Heron, Red-shouldered Hawk, Spotted Sandpiper, Belted Kingfisher, Rough-wing Swallow (which nests under the Lake Avenue bridge), and Scarlet Tanager. Winter is a good season to see woodpeckers, and several species visit the feeders at the Nature Center.

Directions—Park in the lot for the Nature Center on Milwaukee Avenue .75 mile southeast of the intersection with River Road. Other nearby spots to park include the Lake Avenue Woods East picnic area on Lake Avenue west of Milwaukee Avenue

Trails—There are many trails throughout the area.

Note—The Nature Center is open from 9:00 A.M. to 4:00 P.M. in the winter. From March 16 to October 31, the center stays open until 5:00 on weekends and 4:30 during the week. A sugar maple festival is held in March, and a honey festival is held in October.

Telephone—(847)824-8360

Dam No. 1 (K2)   Migrants are plentiful along the river in spring and fall. Woodland nesting species include woodpeckers, flycatchers, vireos, and Scarlet Tanagers, while the meadow often supports Eastern Bluebirds, Grasshopper Sparrows, Bobolinks, and Eastern Meadowlarks.

Directions—Turn south into the parking area from Dundee Road on the east side of the river. Drive past the dam to the far end of the lot.

Trails—Walk south along the bridle path for birding along the river, or take the trail to the east for a long loop through the meadow and back to the river near Willow Road.

Note—**Pottawatomi Woods** north of Dundee Road has occasional Yellow-bellied Sapsucker or Tufted Titmouse in winter. Look for Broad-winged Hawks, Scarlet Tangers, Blue-winged Warblers, Golden-winged Warblers, and Ovenbirds in May/June.

Portwine Road (K1)   This road runs through lowland woods and shrubby meadows on the east side of the Des Plaines River. American Woodcock can be seen and heard performing spring courtship flights in the shrubby meadow, and Eastern Screech-Owls and Tufted Titmice inhabit the woods. The owls can be reliably found by reproducing their call (either by tape or orally) after dark, particularly when the trees are bereft of foliage.

Directions—For the shrubby meadow, park along Portwine Road just south of Lake-Cook Road. This area is also known as Camp Dan Beard. For the woods, park along Portwine Road further south near Dundee Road.

Trails—Walk east into the shrubby area for American Woodcock viewing in spring, and walk west into the floodplain forest to search for woodland species. The trail continues west to the river and the Des Plaines Division Trail.

Ned Brown Forest Preserve (J10)

W S S F

This is a 3,700-acre preserve of woods, open fields, and marsh surrounding Busse Lake. **Busse Forest Nature Preserve** is a 440-acre dedicated Illinois Nature Preserve in the northern portion of the preserve. This large site is good for land-bird migration (particularly in spring) but is also worth visiting at other seasons. In winter, scattered conifer tracts may shelter Long-eared Owls and even an occasional Northern Saw-whet Owl. During the summer, grasslands at the north and south ends have impressive numbers of Willow Flycatchers, Sedge Wrens, Savannah and Grasshopper Sparrows, and Bobolinks. Herons and egrets nest in a colony near the northern boundary, and Yellow-throated Vireos, Rose-breasted Grosbeaks, and Scarlet Tanagers nest in the woods.

Directions—For the southern grassland area, lake, and dam, take Arlington Heights Road south from I-90 to Biesterfield Road and go west to Bisner Road. Turn north on Bisner and take the Forest Preserve road to the first parking area. To reach Busse Forest Nature Preserve, turn north off Higgins Road onto the drive .5 mile west of Arlington Heights Road and park in lot Number 2 or 3 and walk west into the woods. A marsh is located on the west side of the woods. For

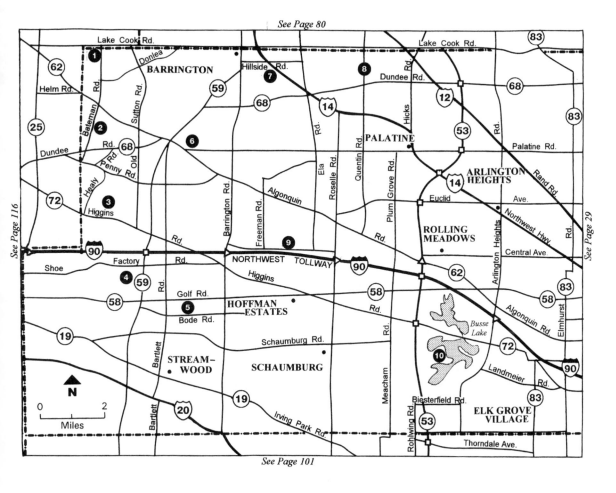

See Page 80

See Page 116

See Page 29

See Page 101

Map J   **Northwestern Cook County**

1.   Spring Lake Nature Preserve *p. 40*

2.   Spring Creek Valley Forest Preserve (Central) *p. 41*

3.   Spring Creek Valley Forest Preserve (South) *p. 41*

4.   Poplar Creek Forest Preserve *p. 40*

5.   Bode Lakes Forest Preserve *p. 40*

6.   Crabtree Nature Center *p. 41*

7.   Baker's Lake Nature Preserve *p. 42*

8.   Deer Grove Forest Preserve *p. 42*

9.   Paul Douglas Forest Preserve *p. 43*

10. Ned Brown Forest Preserve *p. 38*

Busse Lake, turn south from Higgins Road on the road just east of Rte. 53 or the road just west of Arlington Heights Road.

Trails—A paved 7.8 mile bicycle trail circles through the preserve's meadows and forests and offers many views of Busse Lake. It does not enter the Nature Preserve. The trail can be accessed from many of the parking lots. From the lot off of Bisner Road, walk south on the bike path and west to the grassy mound for grassland species.

Poplar Creek Forest Preserve (J4)

This large area of grasslands, woods, and wetlands extends from I-90 south to Schaumburg Road. **Shoe Factory Road Nature Preserve**, a high quality gravel hill prairie, is part of a 600-acre prairie re-creation project in the northwest part of the preserve. The grasslands in the western area are visited by Northern Shrikes in the winter and Short-eared Owls in migration. Nesting species have included Sedge Wren, Eastern Bluebird, Field Sparrow, Vesper Sparrow, Savannah Sparrow, Grasshopper Sparrow, Dickcissel, and Bobolink. Pied-billed Grebes, Wilson's Phalaropes, and Orchard Orioles have been seen from the dike. **Bode Lakes** (J5) are in the southeast part of the area. The woods and savanna adjacent to the lakes can be very good for vireos, warblers, flycatchers, and cuckoos in migration. In summer, the nearby grassy areas support Savannah Sparrows, Bobolinks, Eastern Meadowlarks, and an occasional Northern Harrier. The wetland at Bode and Barrington Road attracts migrating waterfowl.

Directions—From I-90, take Rte. 59 south about .5 mile south of Shoe Factory Road to the lot on the west side of the road. For Bode Lake and Poplar Creek, continue south on Rte. 59 to Bode Road. Turn east and go past Bartlett Road; look for the lot on the north side of Bode Road. For the small wetland at the corner of Bode Road and Barrington Road, park in the lot on Bode Road just west of Barrington Road or across the street in the lot for the strip mall.

Trails—At Shoe Factory Road Prairie, walk toward the hill to the north for the open fields. The pond, lake, and dike are west of the parking lot.

Barrington Area

The Cook County Forest Preserve District owns large tracts of land in the northwestern part of the county that reflect the district's success in preserving a wide variety of landscape types. There are lakes, marshes, prairies, forests, conifer plantations, and re-created savannas. From spring through fall, and to a lesser extent in winter, these sites can provide excellent birding.

Spring Creek Valley Forest Preserve (J2)   Within the borders of this large preserve are lakes, marsh, grasslands, and woods. The birds are as varied as the habitat.

Most birders concentrate on the northern and southern portions of the preserve. The northern 560 acres (Lake-Cook Road south to Donlea Road) have been dedicated as **Spring Lake Nature Preserve** (J1). Rough terrain and the absence of developed trails combine to make this one of the least disturbed tracts in the county. The two natural glacial lakes (Mud Lake and Spring Lake), wooded morainal ridges and extensive wetlands will tempt the adventurous birder. In early spring, Common Snipe and American Woodcock perform courtship displays over marshes and brushy areas, respectively. During spring and summer, Mud Lake hosts Double-crested Cormorants, Great Blue Herons, Great Egrets, and Black-crowned Night-Herons. Probable nesting species include Least Bittern and Marsh Wren in the marsh; Sedge Wren in the sedge meadow east of Bateman Road; Blue-winged Warbler in the brushy old field areas; Yellow-billed Cuckoo and Rose-breasted Grosbeak at the edges of the woods; and Veery, Ovenbird, and Scarlet Tanager in the woods. This is one of the best places in northern Illinois to see summering (and probably nesting) Alder Flycatchers, which inhabit the brushy borders of marsh, prairie and sedge meadow east of Bateman Road.

The southern portion of the preserve (Dundee Road to Higgins Road) attracts numerous migrants and a good selection of summering woodland, grassland, and wetland species. Alder Flycatchers, Veeries, Wood Thrushes, Chestnut-sided Warblers, and American Redstarts have summered in the area just north of the bridge on Penny Road. Nearby are fields populated by Bobolinks (numbers vary greatly from year to year) and Eastern Meadowlarks. At the far southern end of the preserve, the district has established a dog training area just north of Higgins Road. The marsh located here often draws summering herons, Soras, Marsh Wrens, and shorebirds. White-eyed Vireos, Bank Swallows, Yellow-breasted Chats, and Orchard Orioles also summer in the vicinity. The Bank Swallows nest in a nearby quarry and chase insects over the fields.

Directions—To access the northern section (J1), turn south from Lake-Cook Road onto Bateman Road. The preserve is on the east side of the road. Only one small parking area located one-eighth of a mile south of Lake-Cook Road on Bateman Road is available. Cars parked elsewhere may be ticketed.

To access the central section (J2), park on the east shoulder of Bateman Road between Algonquin Road (Rte. 62) and Dundee Road (Rte. 68) or in the lot on the south side of Penny Road west of Old Sutton Road.

To access the southern section (J3), park in the lot for the dog training area on the north side of Higgins Road one mile west of Rte. 59 or in the Beverly Lake parking lot further west.

Trails—To bird the northern portion, there are several choices: (1) to reach Spring Lake and Mud Lake, walk east from the parking area through the marsh, prairie, sedge meadow, and a second marsh; (2) for access to the old field and wooded moraines, walk south of the parking area to the old road which leads east to an oak knoll. From the oak knoll, head south-southeast through the old field habitat to the oak-hickory moraines where there are several bridle trails; and (3) to reach the areas east of Spring Creek and the lakes, walk east from Bateman Road on Donlea Road (closed to car traffic), cross the bridge over Spring Creek, and walk north along the east side of Spring Creek or on the numerous bridle trails. Alternatively, walk east on Lake-Cook Road, then south after crossing Spring Creek.

To bird the central portion, walk east on Donlea Road (closed to vehicles) from Bateman Road through wooded and open areas, and walk north or south along bridle trails. A number of bridle trails are also accessible from Bateman Road between Algonquin and Dundee Roads. From the Penny Road Parking lot, walk north through grassy areas and into the woods.

To bird the southern section, walk north from the dog training area toward the wetlands. From the Beverly Lake lot, walk north along the western edge of the lake.

Crabtree Nature Center (J6)   The principal attractions of this former estate are Crabtree Lake and the much smaller Sulky Pond. Over 30 species of water birds have used one or both of these lakes (primarily during March, April, October, and November), including Common Loon, Horned Grebe, Greater White-fronted Goose, American Black Duck, Northern Pintail, Canvasback, and even a Little

Gull once. About the same number of warbler species have been seen in the shrubby and timbered portions of the preserve during migration. Look for Virginia Rail and Sora in the marsh near the parking lot in April and May. A number of grassland species nest on the site, including Sedge Wren, Vesper Sparrow, Savannah Sparrow, and Bobolink. The much less expected Bell's Vireo has also nested in the preserve. When they are present, winter finches frequent the feeders near the Nature Center, and Northern Shrikes seem to prefer the Phantom Prairie trail.

Directions—The entrance to the preserve is on the north side of Palatine Road one mile west of Barrington Road and .5 mile east of Algonquin Road (Rte. 62). The site is open from 8:00 A.M. to 4:30 P.M. (5:00 in summer). Closed Fridays, the Nature Center is open from 9:00 to 4:30 (5:00 on weekends). The low area one-half mile east of the entrance is referred to as Palatine Road Marsh or Crabtree Flowage. At times it has had enough open water to support migrating shorebirds and breeding herons, waterfowl, rails, Black Terns, and Yellow-headed Blackbirds.

Trails—Go to the Exhibit Building to consult the naturalist and bird chart. Trails provide access to Sulky Pond, Phantom Prairie, Bulrush Pond, and Crabtree Lake. Observation blinds overlook Bulrush Pond and Crabtree Lake.

Telephone—(847)381-6592

Baker's Lake Nature Preserve (J7)   A drained peat bog which spewed smoke for years was finally flooded in the 1930's to extinguish the smouldering fire. Baker's Lake (and presumably cleaner air) was the result. In early spring and late fall, thousands of puddle and diving ducks share the 165-acre lake with an even greater number of American Coots. Other species using the lake are loons, grebes, and regular but infrequent American White Pelican and Tundra Swan. Beginning in 1978, the lake's small island became home to one of the region's major heron rookeries. First to colonize the site were Black-crowned Night-Herons; over the years, they have been displaced by Double-crested Cormorants, Great Blue Herons, and Great Egrets. High water levels have killed the trees that formed the island, but artificial structures enable the herons to continue nesting. The surrounding marsh hosts nesting Mute Swans, Pied-billed Grebes, Wood Ducks, Hooded Mergansers, Virginia Rails, and Soras. During migration, look for warblers in the park on Hillside Road west of the lake and shorebirds on exposed mud.

Directions—Park in the lot on Hillside Road immediately west of Rte. 14 on the north side of the lake. For a view from the south, turn west on Cornell Avenue from Barrington Road and park at Ron Beese Park.

Trails—Walk north and east from Ron Beese Park for a good view of the park from the south. From the small lot on Hillside Road view the lake and walk west on Hillside Road to see the woods on the oak-covered hill overlooking the lake.

Owned By—Forest Preserve District of Cook County, Barrington Park District, and Village of Barrington.

Deer Grove Forest Preserve (J8)   Soon after its formation in 1915, the Forest Preserve District of Cook County purchased its first property: the 500 acres of woods that were to become Deer Grove Forest Preserve. Subsequent acquisitions have nearly quadrupled the size of the preserve, and the diversity of its habitat makes it an exciting place to bird year-round. Stands of conifers attract wintering Long-eared and Northern Saw-whet Owls. Land-bird migration is excellent, and shore-birds such as Greater and Lesser Yellowlegs, Semipalmated Plovers, White-rumped Sandpipers, Baird's Sandpipers, and Wilson's Phalaropes linger where there is mud. A wide range of birds have summered here as well: woodland species include Cooper's Hawk, Red-shouldered Hawk, Wood Thrush, Cerulean Warbler, Ovenbird, and Hooded Warbler; grassland species include Eastern Blue-bird, Henslow's Sparrow, Dickcissel, and Bobolink; and marsh birds include Black-crowned Night-Heron, Virginia Rail, and Sedge Wren. When a beaver dam caused water levels to rise a few years ago, Pied-billed Grebes, Common Moorhens, Black Terns, and Yellow-headed Blackbirds nested in the marsh.

Directions—For the open area of grassland and seasonal wetland in the eastern part of the preserve, turn north from Dundee Road onto the drive .5 mile west of Hicks Road and park in the lot on the east side of the model airplane field. For the area of conifers, turn north from Dundee Road on Quentin Road and park at Camp Reinberg. For Deer Grove West, enter across from Camp Reinberg on the west side of Quentin Road. The road splits with the north fork passing a creek and the south fork ending near a marsh. Both forks have many parking ar-eas, picnic areas, shelters, and people. For the quieter woods in the far western part of the preserve, turn north from Dundee Road onto the drive just east of Rte. 14 and park near the end.

Trails—In the eastern section, walk north across the bicycle path to the creek and wetland area. Depending upon water levels and beaver activity, this may be a marsh, a mudflat, or all dried up. Walk west for the grasslands. From Camp Reinberg, walk to the northeast for the conifers. In the wooded western section, walk on hiking trails, but use a compass to keep your bearings.

Paul Douglas Forest Preserve (J9)   Few people bird this large area, but the mixture of grassland, marsh, and woods attracts a broad array of species. Land-bird mi-gration is good in the spring, particularly in the small wooded section near the preserve's eastern boundary. Of greatest interest, however, are the summering field birds: Savannah Sparrows, Grasshopper Sparrows, Henslow's Sparrows, Dickcissels, and Bobolinks. When the marsh west of the parking lot was flooded by beaver activity, it hosted nesting Common Moorhens and Yellow-headed Blackbirds.

Directions—Exit I-90 at Roselle Road north. Turn west at Central Road and travel about .5 mile to the preserve parking lot.

Trails—Go to the end of the parking lot and walk.

Southern Cook County

Black Partridge Forest Preserve (P1)

W S S F

Almost one-fourth of this 300-acre Forest Preserve has been dedicated as an Illinois Nature Preserve to protect the high-quality woodlands and other natural communities that exist here. Lying almost next to the Des Plaines River, the site attracts an excellent variety of warblers, vireos, and other migrants.

Directions—Park in the lot for the Forest Preserve on the north side of Bluff Road approximately one mile west of Lemont Road.

Trails—Trails lead from the parking lot through the woods and along the stream. The best trails are along the stream and the ridge tops to the east.

Owned By—Forest Preserve District of Cook County

Palos Preserves

W S S F

Built by the ebb and flow of the Wisconsin Glacier and then carved by the glacier's surging meltwater, the Palos region presents a rugged landscape rich with forests, lakes, grasslands, and marshes. Recognizing the value of Palos's locally unique scenery, the Forest Preserve District of Cook County made the preservation of this area an early priority and now owns over 14,000 acres. Unless noted otherwise, the following Palos preserves are owned by the Forest Preserve District of Cook County.

Columbia Woods (P2)   The principal features of this preserve are a narrow strip of large trees along the north bank of the Des Plaines River and a small stream that cuts across the preserve about half-way to the end of the drive. Woodland migrants can be common (particularly around the parking lot) and Prothonotary Warblers have summered, but birders know the woods best for the Pileated Woodpecker that is regularly seen here.

Directions—From Willow Springs Road (104th Avenue), turn west .2 mile south of German Church Road just north of the bridge over the road for the Santa Fe Railway. The turnoff looks like it is going nowhere, but a service road turns south immediately parallel to Willow Springs Road and leads to the entrance of the Forest Preserve. The drive heads west into the preserve and ends after .8 mile at a parking lot. At .5 mile, there is a small boat ramp. Drive slowly and park whenever you hear or see birds. Canoes can be launched here to go downstream to Isle a la Cache or taken out here after launching at Stony Ford.

The **I & M Canal Trail** offers more birding in this area. Drive southeast on 104th Avenue over the river and the canal to Archer. Turn northest on Archer to the first street, Market Street. Follow the signs to the parking lot for the trail. Turn left (west) to the dead end and left again into the parking lot. Walk to the canal for wintering ducks or walk west on the trail. A small creek parallels the trail and many fruit-bearing shrubs can be found along the trail. Tufted Titmice and Carolina Wrens are often present. Farther southwest on Archer, park in the

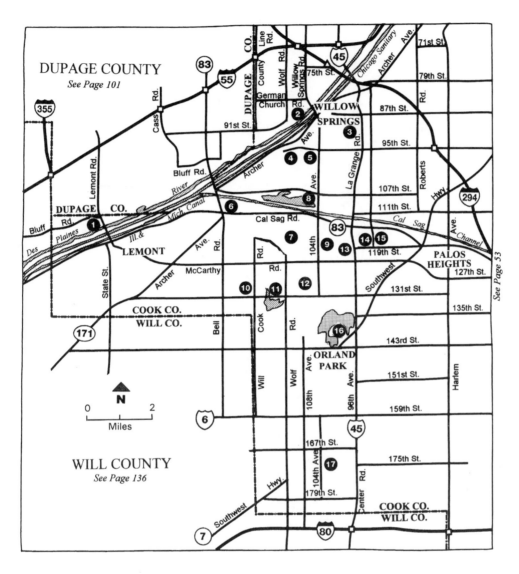

Map P **Southwestern Cook County**

1. Black Partridge Forest Preserve *p. 44*

2. Columbia Woods Forest Preserve *p. 44*

3. Spears Woods Forest Preserve *p. 46*

4. Maple Lake Forest Preserve *p. 47*

5. Long John Slough *p. 47*

6. Camp Sagawau *p. 47*

7. Cap Sauer's Holdings Nature Preserve *p. 48*

8. Saganashkee Slough *p. 48*

9. Cherry Hill Woods Forest Preserve *p. 49*

10. Will-Cook Pond *p. 49*

11. Tampier Lake Forest Preserve *p. 49*

12. Palos West Slough *p. 50*

13. Swallow Cliff Woods Forest Preserve *p. 50*

14. McClaughry Springs Woods Forest Preserve *p. 50*

15. Palos Park Woods Forest Preserve *p. 51*

16. McGinnis Slough Forest Preserve *p. 51*

17. Orland Hills Forest Preserve *p. 52*

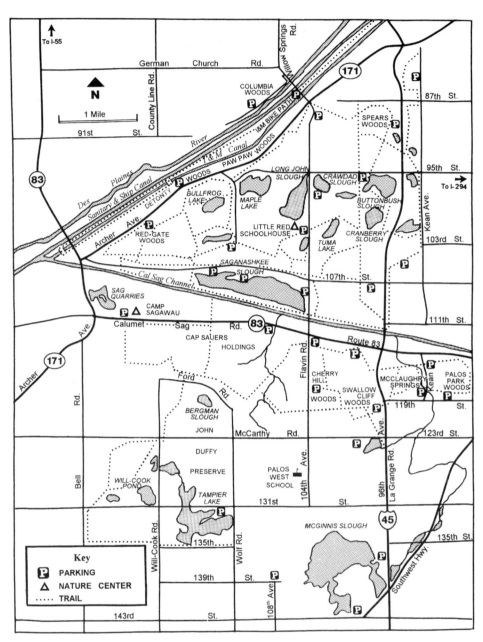

Palos Preserves

lot for **Henry Detonty Woods** and walk west to the trail. White-eyed Vireos may be present in this section.

Spears Woods Forest Preserve (P3)   The mix of woods, grassland, and ponds attracts numerous migrants and an interesting array of breeding species. Tufted Titmice, Blue-gray Gnatcatchers, and Scarlet Tanagers summer in the woods, while Willow Flycatchers, Blue-winged Warblers, Yellow Warblers, Common Yellowthroats, Yellow-breasted Chats, and Field Sparrows summer in the grasslands and shrubby edges.

Directions—From LaGrange Road, pull off into the lot on the west side of the street south of 87th Street and park in the middle of the lot.

Trails—Walk west from the middle of the parking lot to the wide path past the open picnic area. The ponds are to the south (left). By walking north and west, you enter Willow Springs Woods. Some of the trails are also used by equestrians.

Maple Lake (P4)   Situated amidst wooded hills, this deep pond is very good for Common Loons and other water birds in April, October, and November. The woods in the area near Bullfrog Lake to the west have had Blue-winged Warblers, Yellow-breasted Chats, and Orchard Orioles in summer, and a Hooded Warbler was observed in Pulaski Woods south of Maple Lake in June 1997.

Directions—Drive west on 95th Street past 104th Avenue (Willow Springs Road) and park in the lot just west of Long John Slough or turn south after passing Maple Lake and park in the lot on the south side of the lake. For **Paw Paw Woods**, a wooded bluff and floodplain along the Des Plaines River valley, walk north across Archer Avenue.

Long John Slough (P5)   A shallow lake fringed with cattails and willows, it is surrounded by timbered uplands and grassy openings. Overlooking the water is the **Little Red Schoolhouse Nature Center**. The open water and emergent vegetation attract migrating waterfowl, Ospreys, and occasional Bald Eagles, and summering Pied-billed Grebes, herons, Hooded Mergansers, Common Moorhens, swallows, and Marsh Wrens. The woods are excellent for migrant land birds and a good variety of nesters such as Tufted Titmice, Wood Thrushes, and Scarlet Tanagers. In the woodland edges and fields, Eastern Bluebirds, Yellow-breasted Chats, and Orchard Orioles spend the summer, and Ruby-throated Hummingbirds nest in the garden near the Nature Center. Whip-poor-wills and Chuck-wills-widows have nested near the meadow south of the parking lot for the Nature Center. The best time to hear them is at 4:00 A.M. during June.

Directions—Exit I-294 at 95th Street westbound and turn south on 104th Avenue and travel to the lot for the Little Red Schoolhouse on the west side of 104th Avenue. Alternatively, exit I-55 at La Grange Road and go south to 95th Street. The parking lot is open from 8:00 A.M. to 5:00 P.M. (5:30 on weekends). The Nature Center is open from 9:00 A.M. to 5:00 P.M. except Fridays when it is closed.

Trails—From the 95th Street parking lots it is easy to bird from the road. From the Little Red Schoolhouse, it is best to first visit the Nature Center to examine the list of current bird sightings. Several trails begin in the parking lot. Follow the White Oak trail 150 yards south from the Nature Center, then turn west on the horse trail and walk approximately one-half mile to get to the meadow. Please stay on the trails. Guided bird walks are held on some Saturdays from 7:00 to 9:00 A.M. in spring and summer. Call the Nature Center for dates.

Telephone—(708)839-6897

Camp Sagawau (P6)   Sagawau Canyon is one of the few rock canyons in the region and the only one in Cook County. Harboring a locally unusual collection of ferns and other plants, this lovely spot is an Illinois Nature Preserve. In recognition of its fragility, the canyon is off-limits to visitors unless they are with district personnel. Fortunately, frequent opportunities exist for birders to visit the canyon and adjacent woods and fields: the district offers weekly hikes and spring bird walks on Wednesdays, Fridays, Saturdays, and Sundays. Other programs are also offered throughout the year. (Call to learn which programs are scheduled.) Winter birding is enhanced by feeders, which in some years entice wintering Tufted Titmice, Carolina Wrens, Common Redpolls, and Pine Siskins. Spring migration can be impressive here, and birders often tally more than 50 species on a two-hour bird walk. Olive-sided Flycatchers are seen most springs. Unusual migrants have included Bell's Vireo, Worm-eating Warbler, and Louisiana Waterthrush. Black-billed Cuckoo, Blue-gray Gnatcatcher, and Blue-winged Warbler are among the species that have nested.

Directions—Park in the parking area off of Rte. 83 just east of Archer Avenue.

Trails—Walk up the paved drive to the red farmhouse.

Telephone—(630)257-2045

Cap Sauer's Holdings Nature Preserve (P7)   With 1,520 acres, this is the largest nature preserve in Illinois. Perhaps its most prominent feature is Visitation Esker, one of the best examples of this type of glacial landform in Illinois. Avian attractions include migrant passerines and such summering species as Broad-winged Hawk, Eastern Wood-Pewee, Great Crested Flycatcher, Blue-gray Gnatcatcher, Wood Thrush, and Scarlet Tanager. Ospreys have nested at **Bergman Slough** across Ford Road from Cap Sauers.

Directions—Park on the south side of Rte. 83 one half mile west of Willow Springs Road (104th Avenue) for the footpath along the esker (there is an empty signpost near the entrance to the path) or .5 mile further west at the "Take a Walk" sign. You can also park in the lot for Cherry Hill Woods on the east side of 104th Avenue south of Rte. 83, and cross 104th Avenue into Cap Sauers Holdings on the bridle trail one half mile south of the parking lot. Alternatively, you can drive west from 104th Avenue on McCarthy Road for one mile, turn north on Wolf Road, and park after the road turns west (Ford Road). Cap Sauer's Holdings is north of the road, and Bergman Slough is on the south side of Ford Road.

Trails—Take any path or horse trail toward the interior. The area is isolated and extensive; it is easy to get lost here, so go with someone who knows the area or take a compass. A trail parallels Ford Road at the southern edge of the preserve.

Saganashkee Slough (P8)   A large variety of migrant water birds utilize this long and narrow lake which is surrounded by mature woods and brushy areas. Double-crested Cormorants and American Coots are numerous in spring, while concentrations of Common Loons, grebes, and ducks peak in the fall. Bald Eagles are a possibility in any month as they fly over the slough or sit in trees. Ospreys are seen frequently in migration, and Turkey Vultures are seen from spring through fall. Warblers and flycatchers are common during spring migration at the west end of the lake in the woods and brush. An excellent assortment of woodland species have nested in these woods, including Acadian Flycatcher, Yellow-throated Vireo, Veery, Cerulean Warbler, and American Redstart. There is also a colony of Cliff Swallows that nests under the bridge at the eastern edge of the slough on 104th Avenue.

Directions—Park in the lot on the west side of 104th Avenue south of 107th Street or in one of the lots on the south side of 107th Street. One lot is 1.2 miles west of 104th Avenue; the other is .3 mile further west.

Trails—Bird from one of the parking lots or take the trail at the west end of the lake which heads south into mature bottomland woods.

Cherry Hill Woods (P9)   This is a large open preserve with scattered trees, weedy fields, and a pond. A wooded section on its eastern edge connects to **Swallow Cliff Woods**. Migrants are often plentiful in the woods and the pond. Possible breeding birds include Eastern Kingbird, Eastern Bluebird, Field Sparrow, and Orchard Oriole. This area used to attract Alder Flycatchers, Willow Flycatchers, Blue-winged Warblers, Yellow Warblers, Common Yellowthroats, and Yellow-breasted Chats, but the clearing of vegetation for savanna re-creation has altered species composition and abundance.

Directions—Park in the lot on the east side of 104th Street .7 mile south of Rte. 83 and .6 mile north of McCarthy Road.

Trails—Follow the path southeast of the picnic shelter in the grassy area to the east. (The picnic area is heavily used on spring and summer weekends.) The path will circle around to the north and then back to the west to the grassy area. There is another short path which goes south from the grassy area to the small pond. Swallow Cliff Woods preserve is through the woods to the east.

Will-Cook Pond (P10)   A small slough studded with many dead tree stumps, this site is excellent for migrant waterfowl, including Gadwall, American Wigeon, both species of teal, Northern Shoveler, Northern Pintail, and all three mergansers. Eurasian Wigeon have shown up here as well. White-eyed Vireo, Blue-winged Warbler, and Yellow-breasted Chat have been found in spring on the trail south of 123rd Street, and Virginia Rail and Sora have been found in the sloughs along the trail. In fall, the tangled vegetation along the trail attracts Purple Finches and many sparrows.

Directions—Park on the east side of Will-Cook Road just north of 131st Street or on 131st Street at the west end of the slough.

Trails—Waterfowl can be viewed from the car. A trail west of the slough heads north and east from 131st Street through the **John J. Duffy Preserve** to the inter-section of 123rd Street (McCarthy Road) and Will-Cook Road. There are several sloughs and shrubby areas along the trail.

Tampier Lake (P11)   This large fishing lake is on the south side of 131st Street between Will-Cook Road on the west and Wolf Road on the east. Loons, grebes, waterfowl, and Osprey are present in spring and fall. Fishing activity can be heavy, so early morning is the best time to visit.

Directions—Park in the lot on the south side of 131st Street east of Will-Cook Road. **Tampier Slough**, a shallow pond, is one mile north on Will-Cook Road.

Palos West Slough (P12)   The marsh and slough west of the soccer field behind Palos West Elementary School draw a good variety of waterfowl during spring and fall. Nineteen species of waterfowl have been seen on a single day in April. Common Snipe can be heard winnowing in early spring in the tall grasses south of the soccer field, and, if water levels are low, shorebirds feed along the slough's muddy borders in late summer and early fall.

Directions—Exit I-94 at 95th Street and go west to 104th Avenue (Flavin Road). Turn south and proceed four miles to 131st Street. Turn west for .4 mile, then north on Holmes Drive. View the marsh to the north from undeveloped property in the Suffield Woods subdivision south of the school.

Swallow Cliff Woods (P13)   This once heavily wooded site is being thinned as part of savanna re-creation activities. As such, its continued suitability for such forest birds as Tufted Titmice, Veeries, Wood Thrushes, Hooded Warblers, and Scarlet Tanagers is unclear. These species have all summered along the main bridle trail leading away from the parking lot. Eastern Phoebes have nested near the parking lot on the picnic shelter, and Broad-winged Hawks may also nest in the area.

Directions—Turn west into the preserve from LaGrange Road .7 mile south of Rte. 83 and .5 mile north of McCarthy Road. Park in the second lot.

Trails—Birding can be quite good near the stand of White Pines just north of the lot. Walk west on the main bridle trail. By turning north on the first trail, you will end up at the toboggan slides; along the way, the trail parallels a deep ravine. The first trail off the main bridle trail to the south leads to the Laughing Squaw Sloughs and circles around to the east and north back to the parking lot.

McClaughry Springs Woods (P14)   This is a good example of the mesic woods that form in ravines sheltered from fires. Spring-fed Mill Creek meanders through the site, adding to its richness. Winter Wrens and Louisiana Waterthrushes appear in April, the latter occasionally staying to breed. The variety and number of warblers present in these woods in May can be excellent. Nesting species include Wood Duck, Cooper's Hawk, Barred Owl, Ruby-throated Hummingbird, Belted Kingfisher, Eastern Phoebe, Tufted Titmouse, Blue-gray Gnatcatcher, Wood Thrush, Ovenbird, Scarlet Tanager, and Rose-breasted Grosbeak.

Directions—Go east one block from La Grange Road on Rte. 83 (Cal Sag Road) and turn south on Kean Avenue. Park in the lot on the west side of Kean Avenue .4 mile south of Rte. 83 and .2 mile north of 119th Street. **Palos Park Woods** is across the street on the east side of Kean Avenue.

Trails—Mill Creek is just west of the parking lot. Bird the area around the parking lot, the creek, and the bluff beyond the picnic area, then cross the bridge. Birding can be very good at the bridge. From the far side of the bridge, you have the option of walking north along the creek or west up a steep hill. A loop can be made, so walk wherever the birds are singing or where the light is best. If you walk the north trail, you will reach a "T" intersection. To circle back to your car, go up the hill to the west and follow the trail on top of the bluff around to the south. An interesting small path veers off to the west, but keep your eyes open for the trail to the east (about .3 mile from the T intersection) and walk down the steep hill back to the parking lot. If you decide to begin your hike by going up this hill, turn right at each opportunity to end up back at the parking lot.

Palos Park Woods Forest Preserve (P15)   Deciduous woodland and picnic groves across the street from McClaughry Springs provide habitat for resident Great Horned Owls and Tufted Titmice. During migration, this area is excellent for warblers and other woodland birds. A noteworthy collection of species have summered here: Broad-winged Hawk, Red-headed Woodpecker, Carolina Wren, Wood Thrush, American Redstart, Hooded Warbler, and Scarlet Tanager.

Directions—There are two parking areas for Palos Park Woods. For Grove #1, park in the lot on the east side of Kean Avenue south of Cal Sag Road. For Palos Park Woods Grove #2, park in the lot on the north side of 119th Street just east of Kean Avenue. A trail connects the two lots by circling through Paddock Woods to the east. Parking is also available in the lot on 86th Avenue for Paddock Woods.

Trails—If parked at the Grove #2 parking lot, bird the grassy area and trees just west of the lot and take the trail directly west of the grassy area. This trail eventually leads to Kean Avenue and Mill Creek at McClaughry Springs. If the trail is followed to the east, it swings north through Paddock Woods and turns west to the lot for Grove #1 on Kean Avenue.

McGinnis Slough (P16)   On one day in October 1936, the surface of McGinnis Slough (less than a quarter of a square mile in extent) held over 15,000 ducks and 60,000 coots. The open water, marsh and woods that make up the preserve are now surrounded by suburban sprawl, but the slough remains an outstanding birding location. Migrant waterfowl of all kinds arrive in early spring as soon as the ice disappears. The numbers and variety of waterfowl are affected by water levels that fluctuate dramatically from year to year. Occasionally, American Black Ducks, Northern Shovelers, Ring-necked Ducks, and other ducks linger into late spring or summer and may even nest. Other species that have nested include Pied-billed Grebe, Least Bittern, Wood Duck (young are easy to see in July), Common Moorhen, and American Coot. When water levels are low, impressive concentrations of shorebirds and herons appear from July through October. Swallows are abundant from April through July, and careful scanning can often yield all six species. To observe the land bird migrations in May and September, a good location is the west side of the slough where Rusty and Brewer's Blackbirds often occur. Waterfowl using the slough in the fall stay until the ice returns. (Tundra Swans are a good possibility in November.) Bald Eagles are possible anytime from August through December, particularly at the northwest and southwest edges of the slough.

Directions—Three parking areas are used by birders. On the west, park near the intersection of 108th Avenue and 139th Street. On the south, park behind the school on 143rd Street at Union. On the east, park in the lot off of La Grange Road at 137th Street. The entrance to this lot is just .2 mile south of 135th Street, so if you are coming from the north, proceed slowly.

Trails—From the western parking area, walk east through a nice oak woodland and down the hill to the slough. From the southern parking area, walk to the edge of the slough visible from the school. From the LaGrange Road parking lot walk north, follow the path northwest to the island, and walk west.

Orland Hills Forest Preserve (P17)   This large area of fields, shrubby areas, open woodlands, small ponds, marshes, and pine plantations provides interesting birding throughout the year. In winter, Great Horned and Long-eared Owls roost in the pines, and Northern Harriers and Short-eared Owls hunt over the fields. In spring, Alder Flycatchers, Bell's Vireos, or Orchard Orioles may appear in the shrubby areas. Species nesting in the grasslands include Upland Sandpiper, Sedge Wren, Grasshopper Sparrow, Henslow's Sparrow, Dickcissel, Bobolink, and Western Meadowlark. Shorebirds frequent the area in summer and fall if the water is low enough to expose mud flats.

Directions—Turn west from Rte. 45 (La Grange/Mannheim Road) on 167th Street. The preserve is on the south side of the road. Continue to 104th Avenue and turn south. No lots exist, so park on one of the side roads. 104th Avenue has room for several cars south of the pines. Another spot to park is at 172nd and La Grange Road.

Trails—Walk in from one of the gated former roads. Wood ticks are numerous, so wear pants tucked into your socks and use bug spray.

Blue Island Cemeteries (Q2)

W S S F

Six cemeteries with mature trees make an oasis for birds in the vicinity of 115th Street and Kedzie. **Mt. Hope** Cemetery is the largest of these and offers some of the best birding. Evergreen stands often contain wintering Eastern Screech-Owls and Great Horned and Long-eared Owls. Many migrant flycatchers, vireos, and warblers pause in the thickets located in the southeast quadrant of the cemetery. Northern Parula, Black-throated Blue Warbler, and Pine Warbler are among the species birders have found here. This is also a good location from which to see Great Blue Herons, Sandhill Cranes, and a variety of hawks migrating overhead.

Directions—Enter Mt. Hope Cemetery from 115th Street at California Avenue. Go as far south as possible and park straight ahead in the southeast corner or near the building at the west end. The cemetery is open from 8:00 A.M. to 4:20 P.M. Nearby Beverly Cemetery on the east side of Kedzie at 119th Street has a pond and creek. Enter on Kedzie and park at the south end.

Map Q   **Southern Cook County**   (facing page)

1.	Lake Katherine Preserve *p. 220*	9.	127th St. SEPA Station *p. 62*
2.	Blue Island Cemeteries *p. 52*	10.	Van Vlissingen Prairie *p. 57*
3.	Turtlehead Lake *p. 54*	11.	Calumet Park *p. 57*
4.	Burr Oak Woods Forest Preserve *p. 54*	12.	Lake Calumet *p. 57*
5.	Bachelor Grove Forest Preserve *p. 54*	13.	Big Marsh *p. 58*
6.	Elizabeth Conkey Forest Preserve *p. 55*	14.	Indian Ridge Marsh *p. 58*
7.	Vollmer Road Wildlife Marsh *p. 55*	15.	Deadstick Pond *p. 58*
8.	Riverdale Quarry *p. 60*	16.	Torrence Avenue SEPA Station *p. 62*

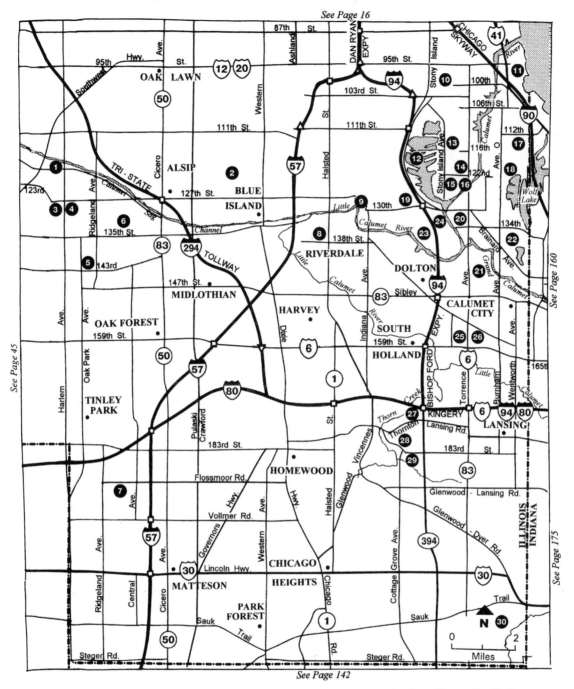

Map Q **Southern Cook County**

17. Eggers Woods Forest Preserve *p. 59*

18. Wolf Lake *p. 59*

19. Calumet Sewage Treatment Plant *p. 60*

20. Hegewisch Marsh *p. 60*

21. Burnham Prairie *p. 61*

22. Powderhorn Lake Forest Preserve *p. 61*

23. Beaubien Woods Forest Preserve *p. 62*

24. O'Brien Lock & Dam *p. 63*

25. Sand Ridge Nature Center *p. 63*

26. Green Lake Woods Forest Preserve *p. 63*

27. Wampum Lake *p. 64*

28. Thornton-Lansing Rd. Nature Preserve *p. 64*

29. Jurgensen Woods Nature Preserve *p. 64*

30. Plum Creek Forest Preserve *p. 65*

Trails—The southern section of Mt. Hope Cemetery is best for birds. Try the hawthorn thickets in the southeast quadrant or the pines and other evergreens in the southwest quadrant.

To bird Beverly Cemetery, walk along the fence and creek to search for migrants in the spring and fall.

Telephone—(708)371-2818

Tinley Creek Preserves

W S S F

In the southwestern part of Cook County, on either side of Harlem Avenue, the Forest Preserve District has significant holdings along Tinley Creek which flows into the Cal-Sag Channel. Within this complex of preserves, birders will find species characteristic of mature woods, fields, and marshes.

Turtlehead Lake (Q3)   This is perhaps the most diverse of the preserves. Evergreens north of the parking lot have sheltered wintering Northern Goshawks and Great Horned Owls. In March and April, American Woodcock make their crepuscular displays to the northwest and south of the lake. Later in spring, vireos and warblers become numerous around the small pond northwest of Turtlehead Lake. The marshes west and south of the lake have Virginia Rails, Soras, and Swamp Sparrows. The drier fields in the same vicinity support nesting Savannah Sparrows, Bobolinks (birders have seen as many as 20 individuals displaying in May), and Eastern Meadowlarks.

Directions—Enter off of Harlem Avenue just south of 135th Street. Drive west and park in the furthest lot (approximately .5 mile). To see the smaller **Arrowhead Lake**, park in the lot on the north side of 135th Street east of Harlem.

Burr Oak Woods (Q4)   This wooded preserve is particularly good for migrant land birds. Unusual species seen here include Kentucky Warbler, Connecticut Warbler, Mourning Warbler, and Hooded Warbler.

Directions—Turn east from Harlem into the preserve south of 135th Street. Bird around the parking lot and walk east to bird the edges of the treeline. **Tinley Creek Woods** preserve is located south of Burr Oak Woods.

Bachelor Grove (Q5)   American Woodcocks use the fields along Ridgeland Avenue to perform their nuptial displays. But of greater interest to birders, the mature woods attract large numbers of migrants, particularly warblers. The following species have summered and/or nested here: Broad-winged Hawk, White-eyed Vireo, Chestnut-sided Warbler, Kentucky Warbler, Hooded Warbler, Yellow-breasted Chat, and Scarlet Tanager. Rough-winged Swallows have also nested near the creek east of the parking lot.

Directions—Park in the lot on the north side of 143rd Street, two blocks east of Harlem.

Trails—Walk on trails through wooded areas near the parking lot and along Justamere Road, an abandoned north-south road west of the parking lot.

Elizabeth Conkey Forest Preserve (Q6)   This preserve on Tinley Creek attracts good numbers of flycatchers, vireos, and warblers during the peak of migration in May. Birding is best when the winds are from the southeast. Nesting species have included Broad-winged Hawk, Yellow-throated Vireo, Blue-gray Gnatcatcher, and Scarlet Tanager.

Directions—Turn north from 135th Street into the preserve east of Ridgeland Avenue and park in the first lot.

Trails—Walk west toward the creek and follow the path north along the creek until you reach the school. Return the same way or walk east to the grassy area and back to your car. If you follow the creek across 135th Street to the south, you enter **Rubio Woods**.

Vollmer Road Wildlife Marsh (Q7)

W S S F

Also referred to as **Bartel Grasslands** (in honor of the late Karl Bartel, longtime birder and naturalist), this 1,900-acre forest preserve has fields, ponds, marshes, scattered evergreens, and deciduous trees. In winter, Northern Harriers, Rough-legged Hawks, and Short-eared Owls work the fields for rodents, and Great Horned Owls and/or Long-eared Owls often roost in pine copses. For several winters, a Great Horned Owl of the white Arctic race was present. The diversity of habitats insures that an equally diverse range of migrants visit the site in the appropriate seasons. Many of the grassland species stay to nest in the numerous fields: Upland Sandpiper south and east of the intersection of Central Avenue and Flossmoor Road; Savannah and Grasshopper Sparrows west of Central and north of Vollmer Road; and Sedge Wren and Henslow's Sparrow in the same area west of Central Avenue and north of Vollmer Road, as well as areas to the north of Flossmoor Road.

Directions—From I-57 southbound, exit at Vollmer Road. Before reaching Vollmer Road, pull over on the shoulder of the exit ramp to scan the pond and marsh west of the highway. At Vollmer Road, turn west and proceed to Central Avenue. Turn north (right) to Flossmoor Road. Turn west on Flossmoor Road to the parking lot on the north side of the road. This lot is open year-round. The lot on Central 1.5 miles north of Vollmer Road is closed in winter. Parking is allowed along Central Avenue, but other roads are posted with "No Parking" signs.

Trails—Few trails exist in this large preserve, but there is a path leading from Central Avenue to the pond visible from the I-57 exit ramp. From the Vollmer Road parking lot, walk north to the field and several areas of pine trees. By parking along Central Avenue just south of 175th Street near the cattail marsh, you can bird two favorite areas, the small field on the west side of Central Avenue near the cattail marsh and the small pond known as "David's Pond" across the street from this field. Deer ticks are plentiful so take precautions and use bug spray.

Calumet Region

W S S F

This area, now associated with rusting steel mills and landfills, once harbored vast acreages of marsh and prairie broken by sand ridges studded with trees. Five shallow lakes provided the most conspicuous relief on the otherwise featureless plain. Meandering south of the lakes and taking their water, the Calumet River was a slow-moving stream that flowed west from morainal uplands in Indiana and doubled back on itself (at Riverdale, Illinois) into parallel channels before emptying into Lake Michigan. Native peoples who inhabited the region thousands of years ago, early European settlers, and

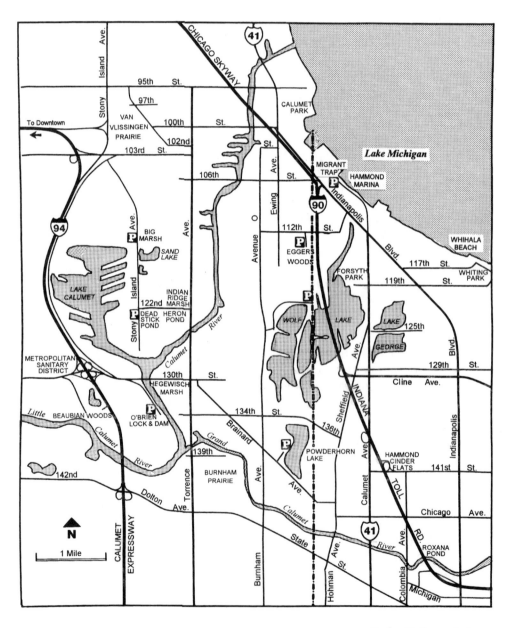

Lake Calumet Area

twentieth century scientists all recognized the incredible biological wealth of this circumscribed region. Old chronicles describe Lake Calumet covered with waterfowl of every kind, and how the sky above darkened when gunfire forced the birds into flight. But 150 years of intensive use filled the lakes (only Calumet and Wolf retain any of their significance), rearranged river courses, and eliminated virtually all of the original wetlands. Most of the marshes that draw birders stand on landfill, and few have any floristic diversity, being dominated by cattails, Common Reed, and the non-native Purple Loosestrife. Still, the birds come, and the Calumet area remains one of the premier birding destinations in the Chicago region.

Van Vlissingen Prairie (Q10)   This 169-acre linear site contains degraded prairie and ephemeral wetlands with small groves of cottonwood trees. The wetland areas draw shorebirds in the spring and fall, and when water levels are high enough, Common Moorhen, American Coot, and a variety of herons are attracted in the summer. Warblers and other migrants can be found in the cottonwood trees in the spring and fall.

Directions—Turn south from 95th Street onto Van Vlissingen Boulevard and park near the corner of 97th Street and Van Vlissingen Boulevard.

Trails—Walk west into prairie via openings along the alley. Van Vlissingen Boulevard is a residential street and the prairie and marsh are on private property.

Calumet Park (Q11)   From this Lake Michigan Beach, birders can observe the few shorebirds that alight and, more importantly, the waterfowl that flock offshore from November through March. Monk Parakeets nest near the Coast Guard station.

Directions—Park in the large lot where 95th Street ends or drive south past the Coast Guard Station to the beach house and park.

Trails—Scan the lake from the 95th Street lot and walk east from the beach house to the lake and beach. Birds seen offshore may count as "Indiana" birds as the state border is approximately 100 yards east of the northern parking lot.

Owned By—Chicago Park District

Lake Calumet (Q12)   Altered in shape and much reduced from its original size, Lake Calumet currently consists of approximately one square mile of water interspersed with finger slips. Large numbers of gulls of various species winter here, and the Chicago region's principal nesting colony of Herring and Ring-billed Gulls is active from late March through early August. The colony, at times exceeding 10,000 breeding pairs, can be seen from the west side of the lake (effective viewing requires a telescope). Herons, waterfowl, raptors, shorebirds, and other birds are present in the appropriate seasons.

Directions—Access is a problem. The northern shore is a golf course, and the eastern shore is fenced for industry. For a view of the west side of the lake and the gull rookery, exit I-94 at 130th Street eastbound. Turn north immediately onto the frontage road and go north for 1.75 miles until you reach Medusa Cement Company. The plant is closed on weekends, holidays, and much of the winter. Walk east to the lake. Further north, views of the lake can be had from the frontage road. This area is isolated, so precautions must be taken for personal and vehicle security. Further development threatens its future as a haven for waterfowl, shorebirds, and gulls.

Owned By—Illinois International Port Authority

Telephone—(773)646-4400

Big Marsh (Q13)   Comprising 290 acres, this is probably the best single birding site in the Calumet area. Water levels fluctuate markedly, which may limit accessibility. Extensive areas of open water attract numerous diving and puddle ducks in early spring and late fall. Until 1997, the state's largest Black-crowned Night-Heron rookery had been situated in the Common Reeds at the southeast corner of the marsh, but recent high water levels may make the area unsuitable. Many other marsh birds have also nested here, including Pied-billed Grebes, American Bitterns, Least Bitterns, Virginia Rails, and Common Moorhens. All six swallows nest in the area and are readily seen as they catch insects over the lake. In late summer and fall, the Big Marsh is a prime location to see Snowy Egrets, Little Blue Herons, and a diverse array of shorebirds including American Avocet, Willet, and Hudsonian Godwit (at least 23 species of shorebirds have been seen here since the mid-1980s). The open area north of the marsh (Railroad Prairie) is a good place to look for raptors, flycatchers, wrens, and sparrows, and the wooded area along Stony Island attracts warblers and other migrant land birds.

Directions—Drive west from Torrence Avenue on 122nd Street, then north on Stony Island Avenue. Drive north one mile and park on the roadside north of the abandoned hazardous waste incinerator. For Railroad Prairie, continue north and park at the open area. Both sites are private property, and access may be denied.

Trails—It used to be possible to walk east across the dike, but high water levels have created a breach in the dike. You may have to bird from "your" side of the dike or walk the edge of the marsh. Sand Lake is on the south side of the dike, Big Marsh is on the north.

Owned By—Waste Management Corporation

Indian Ridge Marsh (Q14)   Made up of 165 acres of marsh and cottonwood trees, Indian Ridge Marsh is bisected by 122nd Street between Torrence Avenue and the railroad tracks. Black-crowned Night-Herons and Great Egrets nest to the north of 122nd Street. If the culvert along the tracks has sufficient water, Pied-billed Grebes, Least Bitterns, and Common Moorhens may nest. Other summering/nesting species include American Woodcock, Willow Flycatcher, Warbling Vireo, Marsh Wren, and Swamp Sparrow.

Directions—Turn west from Torrence onto 122nd Street and park just east of the railroad tracks.

Trails—Walk north along the railroad tracks.

Owned By—Multiple private owners

Deadstick Pond (Q15)   Ducks and other waterfowl congregate in this 80-acre shallow pond and wetland during spring and fall; if the water stays open, many will spend the winter. Pied-billed Grebes, Blue-winged Teal, Northern Shovelers,

King Rails, Common Moorhens, American Coot, and Yellow-headed Blackbirds have summered and/or nested, particularly at the north end. In years where water levels are low enough to expose mudflats, shorebirds may be numerous from July through October. Besides the common species, birders have recorded Wilson's Phalarope, Hudsonian Godwit, Red Knot, White-rumped Sandpiper, Baird's Sandpiper, and Long-billed Dowitcher.

Directions—Drive west on 122nd Street from Torrence Avenue to Stony Island Avenue. Turn south and park along Stony Island Avenue. Be careful of truck traffic from nearby bulk material plants.

Trails—Walk along the berm between the pond and Stony Island Avenue. Cyclone fencing makes viewing a bit difficult.

Owned By—Metropolitan Water Reclamation District of Greater Chicago

Eggers Woods Forest Preserve (Q17)   Two hundred and fifty acres of marsh, sand savanna, slag flats, and bottomland woods, Eggers Woods is an excellent location for many species. Warblers and other passerines are common during spring and fall migration. The marsh has attracted nesting Virginia Rails, Yellow-headed Blackbirds, and other wetland birds. Bell's Vireos, very local in Cook County, have nested in the shrubby area east of the preserve.

Directions—Exit the Calumet Expressway (I-94) at 103rd Street. Go east to Torrence, then south to 106th Street. Turn east on 106th Street to Ewing Avenue, then south to 112th Street. Go east to Avenue "E" and turn south into the preserve.

Trails—Trails circle around and through the woods to the marsh which is on the east side of the preserve.

Owned By—Forest Preserve District of Cook County

Wolf Lake (Q18)   Straddling the Illinois-Indiana border, Wolf Lake is the largest of the Calumet area lakes. The Indiana portion is owned mostly by the City of Hammond which maintains Forsythe Park (Indianapolis Boulevard to 125th Street). The 614-acre William W. Powers Conservation Area takes up most of the Illinois side. From late December through February, Common Mergansers can be found where there is open water, and large numbers of non-native Mute Swans and occasional Tundra Swans gather at the lake's south end. In April and May and again in August and September, look for ducks, Common Moorhens, gulls, and terns in the lake or on the dikes in the lake. Shorebirds often feed long the southern margin of the lake. The woods at the north end of William Powers Conservation Area (called the Wolf Lake Sanctuary) provide excellent habitat for migrating flycatchers, thrushes, vireos, warblers, and other passerines.

Directions—From Illinois, the area can be entered off Avenue "O" approximately .6 mile north of 130th Street. After entering Wolf Lake, turn left and follow the road one mile north and east to the sanctuary. The road is right at the water's edge, so birding for waterfowl can be done from the car. At the end of the drive, turn left and park in the boat trailer parking lot for the Wolf Lake Conservation Area or turn right and drive on the dike down the middle of the lake. The railroad dike divides Indiana and Illinois.

Trails—For woodland passerines, park in the boat trailer parking lot and walk north to the woods. For open water birds, drive south on the dike to the dead end, or bird the lake from the Indiana side.

Note—Access is limited during hunting season from October into early January.

Owned By—Illinois Department of Natural Resources (west side of lake) and City of Hammond (east side of lake)

Calumet Sewage Treatment Plant (Q19)   The sewage settling ponds located here attract shorebirds during the summer and early fall in numbers and variety rivaled by few other places in the Midwest. Over the years it has been a good place to find Ruffs, particularly in early July when some males still have vestiges of their spectacular breeding plumage. But probably the best bird ever seen here was a Sharp-tailed Sandpiper.

Directions—Turn off I-94 onto 130th Street westbound. Turn north at the first drive into the plant. This is a limited access facility. The site manager may require written permission for entry. It is sometimes necessary to park your car and walk the area. Telephone ahead to request permission to enter. No large groups are accommodated, and no more than two cars should enter at one time. Access is limited to weekends.

Owned By—Metropolitan Water Reclamation District of Greater Chicago

Telephone—(773)821-2000

Riverdale Quarry (Q8)   An abandoned quarry with open water and marsh, Riverdale Quarry has become an excellent location for breeding water birds. It became famous among birders in 1981 when 8 adult Eared Grebes produced 12 young. This first (and only) state nesting record demonstrates the adaptability of birds and the occurrence of a marvelous event in an unlikely place. Since then, the cottonwood trees on the far side of the quarry have hosted the largest Double-crested Cormorant colony in the region, as well as smaller numbers of Great Egrets and Great Blue Herons. The marshy border has enabled Soras, Ruddy Ducks, Pied-billed Grebes, and American Coots to nest here as well. The open water is also good for migrant waterfowl of many kinds.

Directions—Park on Halsted near 138th Street and walk .25 mile west to the quarry. There is no parking on 138th Street.

Trails—There is a marsh adjacent to the parking area on the south side of 138th Street. To view the quarry, cross to the north side of 138th Street.

Owned By—Metropolitan Water Reclamation District

Hegewisch Marsh (Q20)   Bordered on the west by the Calumet River, this site contains a large cattail marsh and low woods. The woods provide excellent habitat for migrant passerines (particularly warblers), and the marsh supports breeding Pied-billed Grebe, Blue-winged Teal, Common Moorhen, and Yellow-headed Blackbird.

Directions—Park on Torrence immediately south of 130th Street opposite the gated road to the west.

Trails—Call Waste Management for permission to enter. Enter through the gate on the west side of Torrence and walk west. Watch for the marsh and open water on the left. Walk southeast on the path to the left for an overview of the marsh or continue west to the Calumet River and a view of O'Brien Lock and Dam.

Owned By—Waste Management Corporation

Telephone—(773)646-3099

Burnham Prairie (Q21)   One of the few sites within the Calumet area that contains high quality plant communities, Burnham Prairie is a 175-acre natural area with prairie, bur oak savanna, and marsh. In April at the southern part of the prairie, American Woodcock perform their courtship flights while Sora and Virginia Rails call frequently. Clay-colored Sparrows, Orchard Orioles, and numerous warblers pass through in May. An interesting collection of wetland birds have nested in the marsh including Pied-billed Grebes, Least Bitterns, and Common Moorhens. Burnham Prairie is also an excellent place to see herons. On a single day in early June, seven species foraged in the marsh on the site's western edge.

Directions—Turn east from Torrence on 140th Street and park at the end of the street. This is private property; call for permission to enter. The prairie extends south to 146th Street (State Street). The prairie can be wet in spring, so wear appropriate foot gear.

Owned By—Waste Management Corporation

Telephone—(773)646-3099

Powderhorn Lake Forest Preserve (Q22)   The centerpiece is an artificial lake that was dug out of marsh and prairie as part of a nearby highway construction project. Fortunately, high quality marsh and sand ridges cloaked with prairie and savanna plants have been preserved here. The lake is good for waterfowl in early spring and late fall, and the woods attract a good variety of migrant passerines. The sizable marsh has provided nesting habitat for Pied-billed Grebes, Least Bitterns, Green Herons, Virginia Rails, Soras, and Common Moorhens. Yellow-crowned Night-Herons have nested on private property nearby and are often seen on the preserve. Short-eared Owls are sometimes seen in migration.

Directions—From the intersection of Torrence and 130th St, drive southeast on Brainard. Pass Avenue "O" and turn east into the forest preserve when you see the marsh. For access from the north, turn east from Torrence on 134th Street, turn south on Avenue "K," and park at the end.

Trails—From the end of Avenue "K," walk southeast toward the lake for access to two of the ridges. Three other ridges are accessible from the south.

Owned By—Forest Preserve District of Cook County

Telephone—(773)261-8400

Gull-viewing Sites

W S S F

For many years the Calumet area has been the place where Chicago deposited its garbage. One group of organisms that responded positively to this practice were the gulls in the genus *Larid,* which are now abundant permanent residents. Almost all of the gulls are Herring and Ring-billed, but there is likely to be a smattering of rarer species such as Thayer's, Iceland, Lesser Black-backed, Glaucous, and Great Black-backed Gulls. The following four locations provide opportunities for viewing concentrations of these birds.

Torrence Avenue SEPA Station (Q16)   An excellent spot from which to watch wintering gulls, the parking lot for the Sidestream Elevated Pool Aeration (SEPA) facility is operated by the Metropolitan Water Reclamation District. The facility improves water quality by oxygenating the sluggish waters of the Calumet River.

Directions—Drive north from 130th Street on Torrence. Cross the steel bridge over the Calumet River and turn west immediately into the small drive. Park next to the SEPA facility, but parking is not always available.

Owned By—Metropolitan Water Reclamation District of Greater Chicago

127th Street SEPA Station (Q9)   The SEPA station parking lot provides a view of the Little Calumet River, which at this point remains open year round due to the discharge of warm water by a local steel company. Gulls and other waterfowl can be numerous from December through February. During most years, hundreds of Bonaparte's Gulls gather in the river from October through December.

Directions—Exit I-94 at 130th Street and drive west until road ends at Indiana Avenue. Turn north, then west on 127th Street and park on the north side of the street across from the Water Reclamation District SEPA station for a view of the river.

Trails—View the river from parking area.

Owned By—Metropolitan Water Reclamation District of Greater Chicago

Beaubien Woods Forest Preserve (Q23)   This 250-acre Forest Preserve includes a lake, mature woodlands, grassy meadows, and a boat launching ramp on the Little Calumet River. Railroad tracks divide the preserve into eastern and western portions. The site has its share of spring and fall migrants and grassland nesting species (Field Sparrow, Bobolink, and Eastern Meadowlark), but its principal attraction is the abundance of wintering gulls. From October through March, large numbers of gulls feed at the Dolton landfill across the Little Calumet River and roost on the river itself.

Directions—Access the eastern portion from the southbound lanes of the Calumet Expressway (I-94) south of 130th Street, and the western portion by turning south from 130th Street on the first road west of I-94, Ellis Avenue, then east and south to the boat launch.

Trails—The east side of the preserve has man-made **Flatfoot Lake** and woods. The west portion offers access to the river and boat launch ramp as well as a large grassy meadow. The Dolton landfill may be observed from the boat launch.

Owned By—Forest Preserve District of Cook County

O'Brien Lock & Dam (Q24) Sandwiched between the Calumet River and an active landfill, this site is excellent for gulls. The gulls are attracted by the fresh refuse, and may, therefore, be difficult to locate on Sundays when there is no dumping. Franklin's Gulls have also been seen in the fall during some years. The pond and marshy edge can be good for ducks in spring and fall, and herons during the summer. The site is open from 8:00 A.M. to 5:00 P.M. and has public restrooms.

Directions—Exit east onto 130th Street from the Calumet Expressway (I-94). Pass one stoplight and turn south just before the bridge onto the entrance road to the lock and dam. Continue south to the parking lot.

Trails—The landfill on the west side of the entrance road and the Calumet River and grassy areas on the east side of the entrance road should be checked for gulls and hawks in the winter. The trees and shrubs around the pond south of the main parking lot are good for passerines from May through September.

Owned By—The marsh is owned by the Metropolitan Water Reclamation District of Greater Chicago; the lock and dam is owned by the U.S. Army Corps of Engineers; and the landfill is owned by Waste Management Corporation.

Sand Ridge Nature Center (Q25)

W S S F

Seventy acres of pine plantations, woods, restored prairie, wetlands, and sand ridges surround the Nature Center. Long-eared and Northern Saw-whet Owls occasionally roost in the pines during the winter, and the Nature Center feeders attract various species including Tufted Titmouse. In migration, Lost Beach Trail is a good location to see warblers and other passerines. Nesting species have included Yellow-billed Cuckoo, Red-headed Woodpecker, White-eyed Vireo, Yellow-throated Vireo, Tufted Titmouse, Carolina Wren, Veery, Brown Thrasher, and American Redstart. Hummingbirds are numerous in the fall.

Directions—To get to Sand Ridge Nature Center, Exit I-94 at 159th Street eastbound. Proceed for one mile, turn north on Paxton, and park at the Nature Center on the east side of Paxton Avenue.

Trails—Several trails traverse the Nature Center area. The one-quarter-mile Pines Trail passes through evergreens, oak forest, and prairie restoration. Beyond it, the one-mile Dogwood Trail circles through wetlands and woods and has two bridges over the marsh. North of the visitors' center, half-mile Redwing Trail passes Redwing Pond, and half-mile Lost Beach Trail goes through wet woods and marsh before connecting to a one-mile loop on an ancient beach ridge and sand dune. Boardwalks protect visitors from the wettest places.

Guided walks are held throughout the year. Hours for the lot and trails are from 8:00 A.M. to 5:00 P.M. on weekdays (5:30 on weekends, 4:30 in the winter). The exhibit building is open from 9:00 A.M. to 4:30 P.M. on weekdays (5:30 on weekends, 4:00 during the winter).

Owned By—Forest Preserve District of Cook County

Telephone—(708)868-0606

Note—A forest preserve east of Sand Ridge, **Green Lake Woods** (Q26), can be very good for migrant vireos, thrushes, and warblers. Summering species have included Acadian Flycatcher, Kentucky Warbler, and Yellow-breasted Chat. Drive north of 159th Street on Torrence Avenue and turn east on Michigan City Road. Park on the side of the road before reaching the open area of high voltage power lines. Walk into the preserve on the south side the road before coming to an open area with power lines. Look for the trail into the prairie area to the east. This trail turns south and then west along the ridge which parallels Michigan City Road. As you approach Torrence Avenue, look for the trail which heads south and then east along a wide fire road. This trail veers back north toward your starting point, but in spring, there is often standing water on this trail. Another option is to take the fire road all the way east to the rocky trail which parallels the power lines and walk this trail back north to Michigan City Road. Befitting a low damp woods, this area is loaded with mosquitoes.

Thorn Creek Preserves

W S S F

For several miles south of its intersection with I-294, Thorn Creek flows thorough wet sandy woods that have long been recognized for their botanical significance. Now owned by the Forest Preserve District of Cook County, these sites are heavily visited by both migrants and nesters.

Telephone—(708)868-0606

Wampum Lake (Q27)   Created to provide fill for the Calumet Expressway, Wampum Lake attracts a variety of migrant water birds including Pied-billed Grebe, Blue-winged Teal, Canvasback, Redhead, Ring-necked Duck, Bufflehead, and American Coot. The lake is readily visible from the parking lot. Flycatchers, thrushes, warblers, and other passerines are often common in spring and fall. Ruby-throated Hummingbirds often pause in September at the abundant jewelweed.

Directions—Drive west on Thornton-Lansing Road from Torrence Avenue and turn north into the preserve after crossing over Highway 394.

Trails—Bird the lake from the parking lot or walk to the marshy area and Thorn Creek northwest of the lake.

Thornton-Lansing Road Nature Preserve (Q28)   Also known as **Zander Woods**, this 440-acre wooded state nature preserve protects sandy woods, marsh, and a small bog. Portions of the site were excavated for its sand; the moist depressions now support an unusual community of acidophilic plants. In addition to large numbers of migrants, the preserve attracts such summering birds (some definitely nest) as American Woodcock, Red-headed Woodpecker, Bank Swallow, Tufted Titmouse, Carolina Wren, Blue-gray Gnatcatchers, Veery, Wood Thrush, Ovenbird, and Yellow-breasted Chat.

Directions—Go west of Rte. 394 on Glenwood-Dyer Road to Cottage Grove Avenue, then north one mile to 183rd Street. The preserve is ahead and to the right, but parking is not convenient. Park in the lot for Wampum Lake and walk south.

Trails—A trail heads south from Thornton-Lansing Road into the nature preserve south of Wampum Lake and continues to **Jurgensen Woods** (Q29).

Plum Creek Forest Preserve (Q30)   This large tract of woods, scrub, and large fields lies near the Indiana state line. In the spring, the heavy woods along Plum Creek often harbor a Barred Owl, Acadian Flycatcher or Kentucky Warbler. In spring and summer, look for sparrows, Dickcissels, and Western Meadowlarks in the large fields on the west of Burnham Road. The shrubby areas northeast of Burnham and Katz Corner and along the east side of Plum Creek are good places for nesting White-eyed Vireos, Bell's Vireos, Blue-winged Warblers, and Yellow-breasted Chats. The open areas of this preserve have traditionally been the most reliable spots in Cook County to find Bobwhite. Other nesting birds have included Tufted Titmice, Carolina Wrens, Orchard Orioles, and Henslow's Sparrows.

Directions—From Rte. 30, turn west onto Sauk Trail, then south on Burnham Avenue. Park in the lot at Burnham and Katz Corner Roads for access to the areas west of the creek. For the east side of Plum Creek, continue south to Steger Road, turn east, and park on Steger Road at the bridle trail between the creek and the state line. For the large field area, turn west of Burnham and park along Steger Road.

Trails—On the west side of Plum Creek, trails extend south along the river and northeast to a shrubby area. The bridle trail offers access to the area on the east side of Plum Creek. The large field can be approached from Steger Road or the lot at Burnham and Katz Corner. A footpath from Torrence Avenue on the west also leads to the field.

Note—Nearby **St. Margaret Mercy Health Care Center** (V9) on Rte. 30 at the Illinois/Indiana state line overlooks Plum Creek and has a pond and marsh which attract waterfowl and herons.

Lake County

Lake Border Upland

W S S F

From Highland Park to the Great Lakes Naval base, morainal bluffs rise up to 100 feet above the shores of Lake Michigan. This narrow upland supports an unusual assemblage of natural communities, ranging from harsh eroding bluff-tops to rich forests lining deep ravines. The bluffs offer excellent views of the lake, enabling birders to identify large numbers of wintering and migrating waterfowl. Land birds are common during migration. The ravines provide shelter for at least three species with southern affinities: Tufted Titmouse and Carolina Wren are permanent residents while Louisiana Waterthrush has been known to summer. Winter Wrens and Hermit Thrushes have often wintered.

Rosewood Beach (G17)   Having both a wooded bluff and a beach on Lake Michigan, this site is productive for water birds in fall and winter. Look for Oldsquaw in winter.

Directions—Go east to the end of Roger Williams Avenue. Park and walk down the bluff to the overlook.

Owned By—Park District of Highland Park

Telephone—(847)831-3810

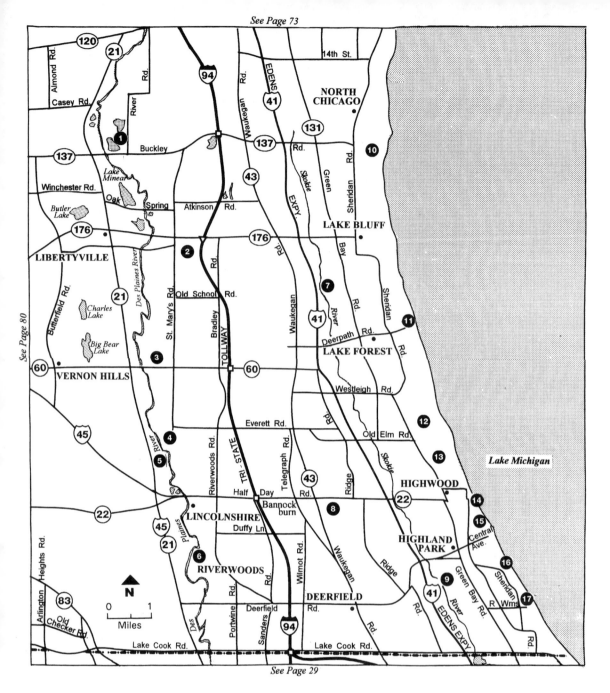

See Page 73

See Page 80

See Page 29

Map G   Southeastern Lake County

1. Independence Grove Forest Preserve *p. 72*

2. Old School Forest Preserve *p. 72*

3. MacArthur Woods Nature Preserve *p. 71*

4. Wright Woods Forest Preserve *p. 71*

5. Half Day Forest Preserve *p. 71*

6. Ryerson Conservation Area *p. 70*

7. Lake Forest Open Lands *p. 68*

8. Prairie Wolf Slough Forest Preserve *p. 69*

9. Foley Park *p. 68*

10. Great Lakes Naval Training Center *p. 68*

11. Lake Forest Beach *p. 67*

12. Lake Forest Nature Preserve *p. 67*

13. Fort Sheridan *p. 67*

14. Moraine Park *p. 67*

15. Park Avenue *p. 67*

16. Millard Park *p. 67*

17. Rosewood Beach *p. 65*

Millard Park (G16)   This wooded ravine can be particularly good for migrant warblers and other land birds.

Directions—From downtown Highland Park, go south on St. John's Avenue/Sheridan Road. One block south of Hazel, turn east on Ravine Drive and follow it through the ravine to Millard Park at the end. Parking is not permitted beyond the gate.

Owned By—Park District of Highland Park

Telephone—(847)831-3810

Park Avenue (G15)   This bluff at the east end of Park Avenue offers a good vantage point for loons, grebes, scoters, and other water birds from fall through spring.

Directions—Turn east from Sheridan Road in downtown Highland Park on Park Avenue which is one block north of Central Avenue. Follow Park Avenue to the end and drive around the water plant for a good view of the lake from your car.

Moraine Park (G14)   This wooded bluff is best in spring and fall for migrant passerines.

Directions—Go east of Edens Expressway on Rte. 22 (Half Day Road). At Green Bay Road, go under the railroad tracks and turn south to Moraine Road, the first street to the east. Follow Moraine Road to Sheridan Road and Moraine Park on the lake.

Owned By—Park District of Highland Park

Telephone—(847)831-3810

Fort Sheridan (G13)   Now closed, Fort Sheridan was originally built in the 1890s to ease the minds of influential north shore residents afraid that the political unrest then rampant in Chicago might sweep their way. Much of the property is being developed into a residential community. Land birds can be numerous in the ravines, waterfowl migrate and/or winter offshore, and raptors migrate overhead.

Directions—The main gate to the property is on Sheridan Road south of Old Elm Road.

Lake Forest Nature Preserve (G12)   Also known as **McCormick Ravine**, this site is the best preserved of the north shore ravines. Interesting throughout the year, it can be particularly good for warblers in spring after a northward flight.

Directions—Park along Sheridan Road or in the small lot east of Sheridan Road at the north edge of Fort Sheridan.

Trails—A trail leads east toward the lake and follows the ridge of the ravine.

Owned By—City of Lake Forest

Telephone—(847)295-9800

Lake Forest Beach (G11)   Waterfowl congregate off the beach (also known as Forest Park and Beach) during fall, early spring, and most winters.

Directions—Follow Deerpath Road to Lake Road and drive down the bluff on the north side of Forest Park. Parking stickers are required in summer.

Trails—Bird from the parking lot.

Owned By—City of Lake Forest

Telephone—(847)295-9800

Great Lakes Naval Training Center (G10)   This is the nation's only training center for naval recruits. Birders focus their attention on the large ravine and the lakefront, with its beach, harbor, and open water of the lake. A number of fall migrants have lingered in the ravine until at least early winter. Waterfowl and shorebirds of many kinds are best seen from September through May. Brant, King Eider, and Black-headed Gull have all been seen here.

Directions—Turn east off Rte. 41 on Rte. 137 (Buckley Road). Turn north on Sheridan Road and look for the entrance gate on the east. Stop for a visitor's card and follow signs to the marina.

Note—It may be necessary to show identification (insurance card, drivers license and registration) to get a pass. Access may be restricted when national security is threatened.

Owned By—United States Navy

Telephone—(847)688-5648

Skokie River Preserves

W S S F

Lake Forest Open Lands (G7)   Also known as **Shaw Woods and Prairie**, much of this site has been dedicated as a state nature preserve to protect the high quality prairie. The mixture of prairie and riparian woods has attracted several noteworthy species during the winter, including Golden-crowned Kinglet, Hermit Thrush, and Purple Finch. But it is unquestionably best for migrant passerines. Almost all locally occurring warblers have been recorded here including Blackpoll Warbler, Worm-eating Warbler, Louisiana Waterthrush, and Connecticut Warbler.

Directions—Turn west onto Laurel Avenue from Green Bay Road and park at the west end of the road.

Trails—Walk the trails to the west and north of the entrance. The areas along the Skokie River are good birding spots, and many birds have been seen near the bridge over the river. Beware of unleashed dogs.

Owned By—Lake Forest Open Lands

Telephone—(847)234-3880

Foley Park (G9)   Consisting of a small woodland pond surrounded by thickets and trees, this park is often excellent for migrant vireos, warblers, and other passerines. A few ducks pause here in migration and both species of night-herons have occurred in summer.

Directions—Turn west from Green Bay Road at the stop light for Bob-O-Link Road opposite Lincoln School south of Deerfield/Central Road. Jog left on Mc-Daniel and continue west on Bob-O-Link. Park before the street turns north into the golf course parking lot.

Trails—Follow the path south along the golf course fence to the small pond. A path circles the pond.

Note—Golf course activity can disrupt birding.

Owned By—Park District of Highland Park

Telephone—(847)831-3810

Prairie Wolf Slough (G8)

This 400-acre newly-created wetland lies along the North Branch of the Chicago River. It is a work in progress, and, to date, ducks and a few species of shorebirds have already availed themselves of the new habitat.

Directions—Park in the lot on the south side of Rte. 22 east of Waukegan Road and the Blackburn Commons shopping center.

Trails—Follow the trail south of the parking lot.

Owned By—Lake County Forest Preserve District

Telephone—(847)367-6640

Des Plaines River Corridor

Some of the best forests in northeastern Illinois lie along the east bank of the Des Plaines River between Deerfield Road and Rte. 176 in Lake County. Classic maple-basswood forests have been able to develop because of the combination of soils that underlie this stretch of river and the absence of fires owing to the protection of the river. There are also fine examples of flatwoods and dry-mesic upland forests, both heavily dominated by oaks. Three tracts of forest have been dedicated as nature preserves. North of Rte. 176, there are more forests, fields, lakes, and an excellent example of a constructed marsh. All of these sites within the corridor are linked by the Des Plaines River Trail which parallels the river for much of its length. Not surprisingly, these preserves are havens for birds year round, particularly during migration and summer. A winter hike along the river, especially when the water is open, will often turn up Brown Creepers, Winter Wrens, Golden-crowned Kinglets, Hermit Thrushes, large flocks of robins, and Yellow-rumped Warblers. During spring and fall, there are days when flycatchers, vireos, thrushes, warblers, and sparrows are thick. Because foliage along the river emerges before it does on the lakefront, outings in late April and early May can produce excellent results. The Wadsworth Road marshes and Sterling Lake are good for such migrant water birds as Common Loon, Horned Grebe, herons, ducks, and shorebirds. The list of birds that have bred in the corridor is also extensive. The southern stretch provides breeding habitat for such forest dwellers as Yellow-crowned Night-Heron, Wood Duck, Red-shouldered Hawk, Broad-winged Hawk, Barred Owl, Acadian Flycatcher, Veery, Cerulean Warbler, Kentucky Warbler, and Hooded Warbler. The fields and wetlands to the north have had summering Pied-billed Grebes, Least Bitterns, Moorhens, Sandhill Cranes, Vesper Sparrows, Grasshopper Sparrows, and Yellow-headed Blackbirds.

The following sites are owned by the Lake County Forest Preserve District.

Telephone—(847)367-6640

Ryerson Conservation Area (G6) Ryerson Woods, particularly during spring migration, is by far the most heavily birded area in the corridor. It has an educational farm, a beautiful Visitors' Center, and a 279-acre dedicated nature preserve. Ryerson's 550 acres of woods and fields attract a wide range of species. The site and the residential areas immediately to the south are the best place in Lake County to find Tufted Titmouse, an increasingly uncommon permanent resident. While never a certainty, one or two Pileated Woodpeckers are often present somewhere on the grounds.

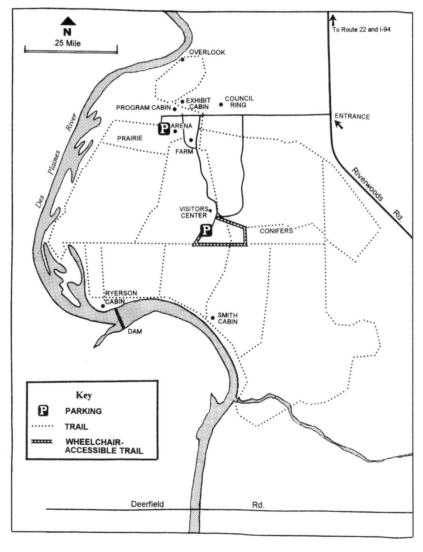

Ryerson Woods

Directions—The entrance to the preserve is on the west side of Riverwoods Road one mile south of Rte. 22 and two miles north of Deerfield Road. Follow the signs to the Visitors' Center and park in the lot. Maps of the trails are available in the Visitors' Center.

Trails—A total of 6.5 miles of trails are available. In winter, visit the stand of spruces or areas along the river; during the rest of the year, any trail may have concentrations of migrants. The River Trail, the Farm Trail, and the North Trail are popular routes.

Telephone—(847)948-7750

Half Day Forest Preserve (G5)   With its picnic shelters, large sports field, fishing lake, and access to the Des Plaines River Trail, this property serves more visitors than most of the other forest preserves in the corridor. But the 200 acres of woods and fields draw their share of birds, particularly during migration. Orchard Orioles, rare in Lake County, have nested near the shelter closest to the lake.

Directions—Park in the lot on the east side of Rte. 21 (Milwaukee Avenue) two miles south of Rte. 60.

Trails—Paths circle the fishing pond and connect to the Des Plaines River Trail. A good trail to follow is the one north of the small bridge over the creek on the north side of the preserve.

Wright Woods Forest Preserve (G4)   This 300-acre preserve features the birds associated with mesic and dry-mesic upland forests. In particular, this preserve is known for its Barred Owls (permanent residents) and nesting Acadian Flycatchers. Both Blue-winged Warblers and Golden-winged Warblers have nested here; even more noteworthy, the rare hybrids produced when these two species interbreed, Lawrence's Warbler and Brewster's Warbler, have occurred as well.

Directions—Exit I-94 at Rte. 60 and go west for 1.5 miles. At St. Mary's Road, turn south. Enter the preserve by going south at the intersection of Everett and St. Mary's Road into the parking lot.

Trails—Four miles of trails loop through the preserve and parallel the Des Plaines River. Be sure to walk the river trails. The Des Plaines River Trail crosses the bridge over the river and passes into Half Day Forest Preserve on the west side of the river.

MacArthur Woods Nature Preserve (G3)   Although less frequently birded than better-known sites in the corridor, this thickly wooded 446-acre preserve has attracted a large selection of migrant and nesting forest birds. Nesters have included Broad-winged Hawk, Barred Owl, both cuckoos, Brown Creeper, Veery, Cerulean Warbler, Black-and-White Warbler, and Canada Warbler.

Directions—No parking lots are close to the preserve. Park in the lot on the west side of the river on the south side of Rte. 60 and walk the trail to the north and across the bridge.

Trails—The Des Plaines River Trail skirts the perimeter of the property. No other trails are developed.

Old School Forest Preserve (G2)   Connected to the river by a narrow strip of land, almost all of this 380-acre recreational preserve is east of St. Mary's Road. It contains areas of shrubs, woods, and fields that attract a variety of migrants, but fewer nesters than the higher quality riparian sites. A feature unique to Old School is the massive sledding hill from which hawks and Sandhill Cranes can be seen during spring and fall migration. Savannah Sparrows and Bobolinks nest in the fields across the street from the main entrance.

Directions—Enter the preserve from St. Mary's Road one-half mile south of Rte. 176. Park in any of the perimeter lots or in the lot by the sledding hill.

Trails—Five miles of trails exist in the preserve.

Independence Grove (G1)   The southern (and largest) part of Independence Grove is a quarry currently being converted into a lake and recreation area. To the north is **St. Francis Woods,** a site recognized by the Illinois Natural Areas Inventory for its high quality forest. Although lightly birded, it has produced a good selection of migrant and nesting woodland species. Two other locations in the vicinity are also worth looking at. The Rte. 137 bridge that spans the Des Plaines River has been colonized by nesting Northern Rough-winged, Cliff, and Barn Swallows, and the low spot in the field west of the river has had shorebirds in the spring.

Directions—To reach the area, go west from I-94 on Rte. 137. The property is on the north side of the road just east of the Des Plaines River.

Trails—The Des Plaines River Trail will traverse the preserve when completed.

Waukegan Lakefront

W S S F

Waukegan Harbor (E11)   The sheltered waters of Lake Michigan and the pilings and piers of the harbor produce a marvelous array of water birds in all seasons except summer. Wintering gulls are common; a careful scanning of the flocks will usually reveal a Thayer's or Glacuous Gull, if not a rarer species. In addition to Oldsquaw, mergansers, and other diving ducks associated with Lake Michigan in winter, Waukegan Harbor offers less common wintering waterfowl, lured here by the bread left by kind-hearted residents. Wood Ducks, American Wigeon, Hooded Mergansers, and American Coots are among the birds that often linger with the Mallards. In flight years, this is an excellent place for Snowy Owls. Two of the area's rarest shorebirds appear here with some regularity in late fall: Purple Sandpipers on rocks and Red Phalaropes in the harbor.

Directions—Drive east of Sheridan Road and the Amstutz Expressway on Grand Avenue (Rte. 132) to Pershing. Turn right on Pershing Road and take the first left (Clayton Street) to the lot north of the Waukegan Yacht Club to view the "old" harbor. The new harbor is south of government pier and can be reached by driving two blocks on Pershing and turning east at the sign. Jog south and east to the parking lot to view the new harbor from your car.

Trails—Bird from either lot or walk out on the jetty in the harbor.

Owned By—The Waukegan Port District

Telephone—(800)400-7547

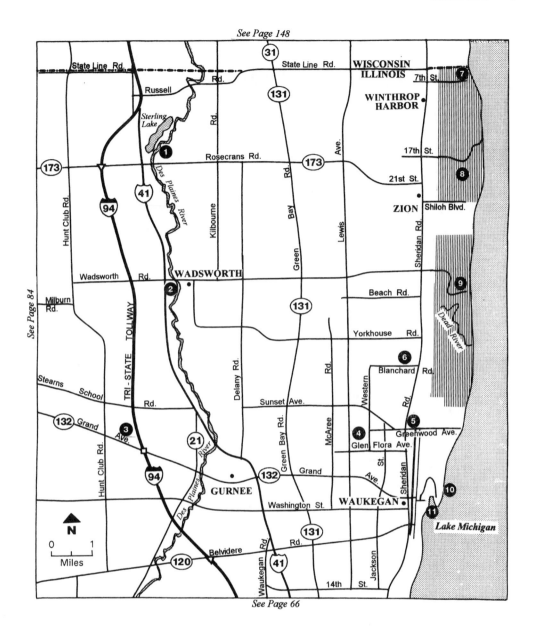

See Page 148

See Page 84

See Page 66

Map E **Northeastern Lake County**

1. Van Patten Woods Forest Preserve *p. 79*

2. Wadsworth Wetlands Demonstration Project *p. 78*

3. Gurnee Mills Ponds *p. 79*

4. Larsen Park *p. 75*

5. Bowen Park *p. 75*

6. Lyons Woods Forest Preserve *p. 75*

7. Spring Bluff Forest Preserve *p. 78*

8. Illinois Beach State Park North Unit *p. 77*

9. Illinois Beach State Park South Unit *p. 76*

10. Waukegan Beach *p. 74*

11. Waukegan Harbor *p. 72*

BOWEN
PARK
BALLFIELD

Greenwood   Ave.   FENCE   **P**   FENCE

COM ED

FISHING
PIER

EXPRESSWAY

Pershing   Rd.

N

.25 Mile

Sheridan Rd.

RR TRACKS

RR TRACKS

FENCE

Lake Michigan

AMSTUTZ

Sea   Horse   Drive

BEACH

BEACH
HOUSE

BEACH PIER

(132)
Grand
Ave.

Clayton
Pl.

Harbor   Pl.

DEPOT

P

P
WATERWORKS

WAUKEGAN YACHT CLUB

NORTH JETTY

Washington

P
SOUTH
HARBOR

RR TRACKS

Belvedere

Key

|||||||| FENCE
▬▬ RR TRACKS
P PARKING

Waukegan Harbor Area

Waukegan Beach (E10)   Situated between the Commonwealth Edison power plant on the north and Waukegan Harbor on the south, this stretch of beach is like no other in Illinois. Although significantly altered from its original state, it includes a mixture of sandy beach and inlets, grassy foredunes, marsh, woods, and ponds. In winter, it often hosts a Snowy Owl or Northern Shrike and roosting gulls. The power plant cooling pond, visible from the fishing pier at the end of Greenwood Avenue, often holds wintering puddle ducks, Canvasbacks, Redheads, and even an occasional Harlequin Duck or scoter. During migration, raptors and land birds can be excellent. But what makes this beach such an important destination for birders is the potential for shorebirds. From late April though September, Willets, Ruddy Turnstones, Red Knots, Sanderlings, and peeps often stop to feed. The beach is among the best sites in the Chicago region for the less common Piping Plover, American Avocet, and Whimbrel. In summer, Ring-billed Gulls nest west of the beach, and the only Common Tern nesting colony in Illinois is at the north end of the beach. Although not known to breed here, Caspian Terns can be common. Further augmented by the occasional summering Laughing, Franklin's, Bonaparte's, and Glaucous Gull, the gull and tern activity gives the beach a maritime feeling.

Directions—Drive east of Sheridan Road and the Amstutz Expressway on Grand Avenue and south on Pershing Road. Turn east on Clayton Street, then north on Sea Horse Road and follow it past the parking lot for the beach house to the lot for OMC Dock #5 at the end. A parking fee is charged between Memorial Day and Labor Day. The lake can also be viewed from the end of Greenwood Avenue two miles to the north, but the beach to the south is not accessible from this point. From the end of Greenwood Avenue, walk out on the fishing pier to see the Commonwealth Edison cooling pond or look south through the fence at a smaller pond.

Trails—Check the pines behind the beach house for land bird migrants. Walk out on the north jetty to check for Purple Sandpipers (rare) in fall and to view the harbor. Walk 1.5 miles north along the beach as far as the power plant where a pond often forms on the beach. You will have to ford a stream or two on your way north. If you can pull yourself from the open beach, bird the woods and marshes on the western edge.

Bowen Park (E5)   This city park with a wooded ravine can be excellent during migration for flycatchers, warblers, and other passerines. The ball diamond is a good place to observe migrating hawks.

Directions—Turn east from Sheridan Road into the parking lot just north of Greenwood Avenue and park in the northeast lot, next to the Jack Benny Center.

Trails—Walk down into the ravine and to the south.

Note—Larsen Park (E4), west of Bowen Park on Western Avenue, south of Greenwood, can be very good in spring for vireos, thrushes, and warblers.

Illinois Dunes

A narrow strip of the Chicago Lake Plain extends from Waukegan to northern Racine. Up to the Illinois/Wisconsin boundary, this area of dunes used to be called the Waukegan Moorlands. The southern portion includes Waukegan Harbor and the beach area just discussed. Fortunately, much of the northern area is protected as state park or Forest Preserve. The great ecological value of this section of lake plain was confirmed by the Illinois Natural Areas Inventory, which found that more high quality natural communities exist here than any other area in the state. From May to September it is a botanical and entomological wonderland. It is also, of course, a superb place to bird.

Lyons Woods (E6)   On the upland just west of the lake plain, this 264-acre preserve consists of deciduous woods, grassland (some of which is native prairie), and mature stands of conifers in a former nursery. During the winter doldrums, the evergreens here can often yield such exciting finds as Long-eared Owls, White-

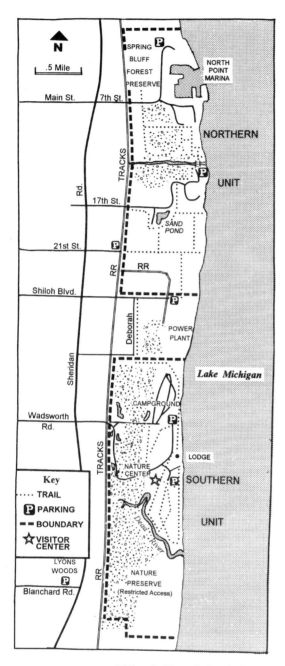

Illinois Beach State Park

winged Crossbills, Common Redpolls, and Pine Siskins. Sometimes a Northern Shrike spends the season in the more open areas. Eastern Bluebirds have wintered here, as has a Townsend's Solitaire on one known occasion. The site is heavily visited by migrant land birds, some of which have summered and/or nested. These include Red-breasted Nuthatch, Golden-crowned Kinglet, Chestnut-sided Warbler, Hooded Warbler, and Yellow-breasted Chat.

Directions—Take Blanchard Road west of Sheridan. Park in the lot for the preserve on the north side or further west on North Avenue north of Blanchard Road.

Trails—Take the trail north from the preserve parking lot. The pines are on the east of the trail before it crosses a meadow and reaches an east-west trail. Take the trail toward the west through the oak savanna, cross North Avenue, and enter the prairie. The total length of the trail is three miles.

Owned By—Lake County Forest Preserve District

Telephone—(847)367-6640

Illinois Beach State Park South Unit (E9) The heart of the Illinois dunes is this 3,000 acre tract of beach, low dunes, swales, oak-covered ridges, marsh, ponds, river and a 120-year old planting of European evergreens. The south unit was the first portion of the dunes to be made a state park and is the most popular of all Illinois state parks, drawing over 2 million visitors a year. The park headquarters, lodge, and Nature Center are located here. To protect the park's natural features, the southern two-thirds have been designated a state nature preserve, the very first established under Illinois's landmark Natural Areas Preservation Act. Winter birds can include waterfowl offshore, various finches in the conifers, Northern Shrikes along the Dead River, and Snow Buntings near the lake. Spring land bird migration starts with Eastern Phoebes, Eastern Bluebirds, Hermit Thrushes, and Fox Sparrows in early April. By late May to early June, the end is signaled by the appearances of Olive-sided Flycatchers, Yellow-bellied Flycatchers, and Connecticut Warblers. Species that have nested are as varied as the habitat and include Pied-billed Grebe, Least Bittern, Black-crowned and Yellow-crowned Night-Heron, King Rail, Upland Sandpiper, Whip-poor-will (easily heard at night), Golden-winged Warbler, Mourning Warbler, Vesper Sparrow, Henslow's Sparrow,

Western Meadowlark, and Brewer's Blackbird. Eastern Bluebirds nest in the area of Black Oaks near the Nature Center. The fall land bird migration extends from late August to November, and is just as rewarding as the spring. But for many birders, the highlight of an Illinois Beach fall is its hawk migration. Beginning in the middle of September, days with winds having a westerly component bring Ospreys, Sharp-shinned Hawks, Cooper's Hawks, Broad-winged Hawks, Merlins, and Peregrine Falcons south through the park in substantial numbers. This may be the best place in Illinois to see multiple Merlins and Peregrines, for their flight paths tend to hug the lakeshore. October brings Northern Harriers, Red-tailed Hawks, and more Merlins. Finally, in November, when it becomes too cold for all but the most dedicated hawk watchers, Red-shouldered Hawks, Rough-legged Hawks, and a few eagles (of both species) drift through. The south unit of Illinois Beach also boasts an amazing list of rarities, including Harris's Hawk, Prairie Falcon, Whooping Crane, and Scrub Jay.

Directions—Enter the park from Wadsworth Road. The deciduous woods south and east of the railroad tracks can be very good for passerines during migration. For the Dead River, Nature Center, and observation platform, continue south and east and park in the lot near the Nature Center. For an overview of the lake, continue to the lot for the lodge or the beach. The park headquarters is north of the lodge.

Trails—Several trails head south and east from the Nature Center. The Dead River Trail parallels the Dead River and ends at the beach. The Loop Trail circles through the woods and dunes and has an observation platform overlooking the beach. The Beach Trail heads east from the Nature Center and has a boardwalk across a wet swale. A good walk can be down the beach and back up through the woods or vice-versa depending upon the light. To observe raptor migrations, go to the observation platform near the lake, stand on the Dead River Trail near the beach, or sit on the deck at the Nature Center. Access to the area south of the Dead River is not allowed without a permit, and it is no longer possible to reach the area from Greenwood Avenue further south. Call ahead to request a day permit. Be sure to check the small pond behind the lodge on the south side. In the lodge, there are a restaurant and a shop with maps of the park.

Note—The Nature Center is open from 9:00 A.M. to 12:00 noon and 1:00 P.M. to 3:00 P.M.

Owned By—Illinois Department of Natural Resources

Telephone—(847)662-4811

Illinois Beach State Park North Unit (E8) With 1,160 acres, the north unit of the state park is both smaller and less diverse than the south. A large part of the property has been dedicated as a state nature preserve in recognition of its undisturbed prairie, wetlands, and savanna-clad ridges. In winter, the grasslands usually have raptors and a Northern Shrike. Land bird migration is somewhat limited by the paucity of trees, but sparrows and ground feeding warblers such as Canada and Mourning utilize the shrubs. Accipiters and falcons often fly close to the shore in September and October. Most birders, however, concentrate on Lake Michigan and the beach, where they seek water birds, gulls, and shorebirds. They also check Sand Lake, which over the years has attracted a good collection of waterfowl including scoters and Tundra Swans. A crepuscular visit to the North Dunes Nature Preserve in late April or May will almost certainly yield performances by that marvelous trio of Common Snipe, American Woodcock, and Whip-poor-will.

Directions—Turn east from Sheridan Road on Shiloh Boulevard in Zion for the North Dunes Nature Preserve and the lakefront at the Zion Nuclear Power plant. Further north, turn east from Sheridan Road on 21st Street and park at the end near the Nature Preserve or turn east on 17th Street to the area known as **Camp Logan** and **Sand Pond**.

Trails—Several trails crisscross the area and go from one parking area to the others. From the Shiloh Boulevard lot, walk north along the beach or along the multipurpose trail to Hosah Park, a Zion park with an elevated viewing platform on the beach, or continue .5 mile north on the trail to the continuation of 21st Street and the North Dunes Nature Preserve.

Owned By—Illinois Department of Natural Resources

Telephone—(847)622-4811

Spring Bluff Forest Preserve (E7) Lightly-birded Spring Bluff Forest Preserve consists of high quality savanna, prairie and wetlands that have provided nesting habitat for such birds as Least Bittern, three species of rails, Common Snipe, American Woodcock, and Sedge Wren. More birding effort, however, has focused on **North Point Marina**. From fall through early spring, it is a major gathering place for gulls, including such rarities as Iceland and Lesser Black-backed. Ducks rest on the marina's calm waters and an occasional Snowy Owl perches on a piling.

Directions—Go east of Sheridan Road on 7th Street to the lake. Park south of Skipper Bud's boat storage buildings and walk out on the drive to view the lake from the picnic area. A sandy beach can be viewed north of the boat launch ramps north of Skipper Bud's. To view the south end of the marina, take the drive south of 7th Street. Access to Spring Bluff is from the parking lot north of the marina.

Trails—Spring Bluff Forest Preserve is fenced off from the lot, but access is possible by walking north of the large boat storage building on the dirt path and turning west on the small trail before reaching Chiwaukee Prairie and Prairie Harbor in Wisconsin.

Owned By—The Forest Preserve is owned by Lake County Forest Preserve District; the marina is managed by the Illinois Department of Conservation.

Wadsworth Wetlands Demonstration Project (E2) Once an area of quarries, old fields, and scrubby woods, much of this 450-acre site has been converted into a series of marshes where nesting Common Moorhens, Yellow-headed Blackbirds, and other wetland birds can be easily observed. In late summer, large numbers of herons gather here (including Snowy Egrets and Little Blue Herons on occasion), and shorebirds occur when water levels are low. Migrant land birds can be common in the wooded sections along the river; the oaks located approximately one block south of Wadsworth Road are particularly good.

Directions—Go east of Rte. 41 on Wadsworth Road and park in the lot on the south side of the road just east of the Des Plaines River and west of the railroad tracks.

Trails—Walk the 1.5-mile path which makes an oval loop through the preserve. The ponds and marsh are on the west side of the preserve. The woods are on the east side and down the middle of the preserve, and shrubby areas are to the south.

Van Patten Woods (E1)   Both the woods on the east side of the river and the open waters of 74-acre **Sterling Lake** attract a wide variety of migrants. When lake levels drop sufficiently to expose muddy edges, shorebirds augment the mix. During the summer, Grasshopper Sparrows and Bobolinks nest in the grasslands.

Directions—Enter the preserve on the north side of Rte. 173, just east of Rte. 41. Turn on the first road to the east and park in the first lot for the oak woods or continue further to the pine plantings. To view the lake and the grasslands on the west side of the preserve, drive straight ahead after entering the preserve and cross the bridge over the river. The north side of the preserve and model airplane flying field can be entered from Russell Road, off of Rte. 41.

Trails—One- and two-mile trails loop through the oaks and pine plantings east of the river and circle around the two basins of Sterling Lake. The Des Plaines River Trail begins on the north side of the preserve and continues south between the river and Sterling Lake. The trail continues south through the **Wadsworth Prairie** and Des Plaines Wetlands Project.

Gurnee Mills Ponds (E3)

While your loved ones are shopping at this huge mall, sneak off to the marsh and shallow ponds on the east side of the complex. Herons, ducks, shorebirds, and Forster's Terns can be found in season.

W S S F

Directions—Exit I-94 at Grand Avenue (Rte. 132) westbound. Enter the lot for the mall and go around to the east side toward I-94 and park near the gazebo. Ask security personnel inside the mall for permission to walk around the nature area.

Trails—The gazebo affords a good overlook.

Buffalo Creek Forest Preserve (F8)

For the sake of flood control, Buffalo Creek was dammed to create a reservoir. The lake, a section of free-flowing creek fringed with cottonwoods, and the surrounding old fields became this 396-acre preserve. The three principal habitat types are all attractive to migrants, and the fields have supported nesting Vesper Sparrows, Grasshopper Sparrows, Dickcissels, and Bobolinks.

W S S F

Directions—Turn west from Arlington Heights Road onto Checker Drive. The entrance to the preserve is on left.

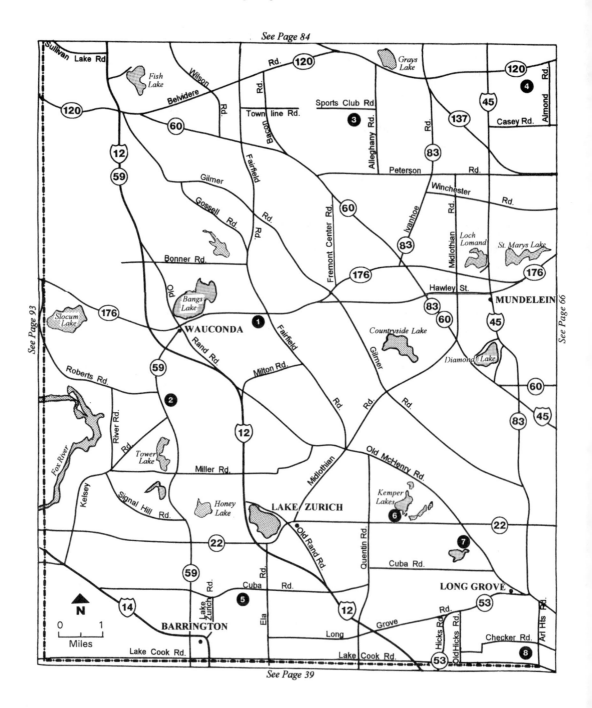

See Page 84

See Page 39

Map F   **Southwestern Lake County**

1. Lakewood Forest Preserve *p. 82*
2. Summerhill Estates *p. 82*
3. Campbell Airport *p. 83*
4. Almond Marsh Forest Preserve *p. 83*
5. Cuba Marsh Forest Preserve *p. 81*
6. Kemper Lakes *p. 81*
7. Reed - Turner Woodland *p. 81*
8. Buffalo Creek Forest Preserve *p. 79*

Trails—Walk along the bike paths that wend through the preserve.

Owned By—Lake County Forest Preserve

Telephone—(847)367-6640

Note—**Buffalo Creek Nature Preserve** east of Arlington Heights Road attracts warblers in migration.

Reed-Turner Woodland (F7)

These 32 acres of mostly wooded hills contain a winding creek and a portion of a small lake. Spring is the time to visit this preserve, both for the wildflowers and the woodland migrants.

Directions—Look for the sign on the west side of Old McHenry Road .25 mile south of Rte. 22 and park in the lot.

Trails—Follow the marked trail out of the parking lot and across Indian Creek. At the "T" intersection, a right turn will take you to the west end of Salem Lake; a left turn will take you to the east end of the lake, to a small prairie, through the woods, and back across the river and another bridge to the parking lot.

Owned By—Long Grove Park District

Telephone—(847)438-7230

Kemper Lakes (F6)

The lakes associated with this well-known golf course attract a great variety of migrant waterfowl including Greater White-fronted Goose (best seen in very early spring), Northern Shoveler, Canvasback, Redhead, Common Goldeneye, all three mergansers, and Ruddy Duck. Some of these birds spend the winter, as the water usually remains open.

Directions—Drive west on Rte. 22 and enter the property on the north side of the road one mile west of Old McHenry Road. Access may be restricted at times.

Owned By—Kemper Insurance Company

Telephone—(847)320-2000

Cuba Marsh (F5)

The marsh that curls along the east side of the preserve possesses a combination of open water and emergent vegetation that maximizes avian diversity. Many ducks use the marsh in migration, and herons forage in the summer. Displaying woodcocks are easily found in early spring. Wetland species nesting here include Pied-billed Grebe, Least Bittern, Blue-winged Teal, Virginia Rail, Sora, Common Moorhen, American Coot, and Marsh Wren. Another section of open water is on the west side of the preserve and is reached via active railroad tracks. This 780-acre preserve also has upland areas of grass, shrubs, and woods that are productive for migrants and nesters. Recent breeding surveys have located Acadian Flycatchers, Alder Flycatchers, Yellow-throated Vireos, Blue-winged Warblers, American Redstarts, Grasshopper Sparrows, and Bobolinks.

Directions—Park in the lot on the south side of Cuba Road just west of Ela Road.

Trails—Two miles of trails exist in the preserve between Cuba Road and Ela Road. The fields west of the tracks are good for grassland birds, but parking is restricted along Zurich Road.

Owned By—Lake County Forest Preserve District

Telephone—(847)367-6640

Lakewood Forest Preserve (F1)

W S S F

With 1,850 acres, Lakewood is the largest of the Lake County Forest Preserves. Much of it is in cultivation, but there are lakes, marshes, woods, and old fields. The wooded sections are good during migration, and numerous birds nest throughout the preserve. Highlights of the 1998 breeding census include Sandhill Cranes, Black Terns (on the lake next to the first parking lot), Wood Thrushes, Blue-winged Warblers, Ovenbirds, Yellow-breasted Chats, and Grasshopper Sparrows.

Directions—From Rte. 176, turn south onto Fairfield Road, and then southwest onto Ivanhoe Road which bisects the preserve. Turn right into the first parking lot to access the small marshy lake. Continuing on Ivanhoe, turn left to reach the horse trails. Still continuing on Ivanhoe, turn right toward the museum and left on the gravel road to Shelter 11 where the nature trail begins.

Trails—There are nine miles of hiking trails in the preserve. The horse trails and the nature trail are recommended.

Telephone—(847)367-6640

Owned By—Lake County Forest Preserve District

Potholes and Marshes

W S S F

Scattered across the western half of Lake County are wet areas that draw a wide assortment of water birds. Some of these areas are merely flooded depressions whose suitability for birds is as short-lived as the standing water. Others are substantial open water marshes that stay permanently wet. From year to year, precipitation levels and development help determine which existing sites are lost and which new ones are created or made accessible. The six wetlands included here have all been productive over a period of years.

Summerhill Estates (F2) During migration, Gadwall, Green-winged Teal, Redheads, Ring-necked Ducks, and other ducks loiter on the shallow lake within this private subdivision. In late spring and early summer, Common Moorhens and Black Terns are often seen. When conditions allow, shorebirds and herons forage on the edges.

Directions—North of Rte. 176, turn east of Fairfield Road on Old Oak Drive.

Trails—Although it is still possible to bird from the road, construction of more homes may soon obstruct views of the water.

Almond Marsh (F4)   An excellent mix of open water, flooded trees, and cattails, this marsh is good for both migrating and nesting water birds. Pied-billed Grebes, Virginia Rails, Common Moorhens, and Yellow-headed Blackbirds breed here, as do Great Blue Herons which recently established a small colony in the dead trees at the edge of the marsh. The upland woods can produce warblers and other migrant passerines.

Directions—From Belvidere Road (Rte. 120), turn south onto Almond Road east of Rte. 45. Look for the sign for Almond Marsh on the west side of the road. This drive is also the entrance to the Lake County Forest Preserve Planning office. Park in the lot at the end of the drive.

Trails—The marsh is easily visible from the parking lot.

Note—A gate across the drive is locked on weekends. To make arrangements to enter, call the Lake County Forest Preserve offices.

Owned By—Lake County Forest Preserve District

Telephone—(847)367-6640

Campbell Airport Area (F3)   When the large pothole just west of Alleghany Road is wet in April and early May, it usually attracts ducks, shorebirds, and American Pipits. In the spring of 1998, the pothole was dry, but a pond north of Sports Club Drive was productive. Horned Larks are reliable in spring and usually nest here.

Directions—Drive west from Alleghany Road on Sports Club Drive north of Peterson Road. The pond is on the south side of the road. Bird from the road as this is private property.

Trails—Look north and south from the road for flooded areas.

Nippersink Marsh (D14)   During migration this marsh with open water can be very good for such ducks as Gadwall, American Wigeon, Northern Shoveler, Bufflehead, and Ruddy. In summer, look and listen for Common Moorhen and Yellow-headed Blackbirds.

Directions—Turn east of Fairfield Road on Nippersink Road north of Rte. 120. The marsh is on the south side of the street across from Village Elementary School. Park in the lot for the school or nearby church.

Red-Wing Slough (D5)   This is one of the outstanding wetlands in northern Illinois, but access is largely restricted to what can be seen from both sides of Rte. 173. Waterfowl of many species use the large cattail marsh and shallow lake in migration. Least Bitterns, Common Moorhens, Sandhill Cranes, Forster's Terns, Black Terns, and Yellow-headed Blackbirds have all nested in areas north of the road.

Directions—Park along Highway 173 1.5 miles west of Rte. 45 to view the preserve. There is no parking lot. Although the shoulder is wide, traffic is heavy, so exercise caution.

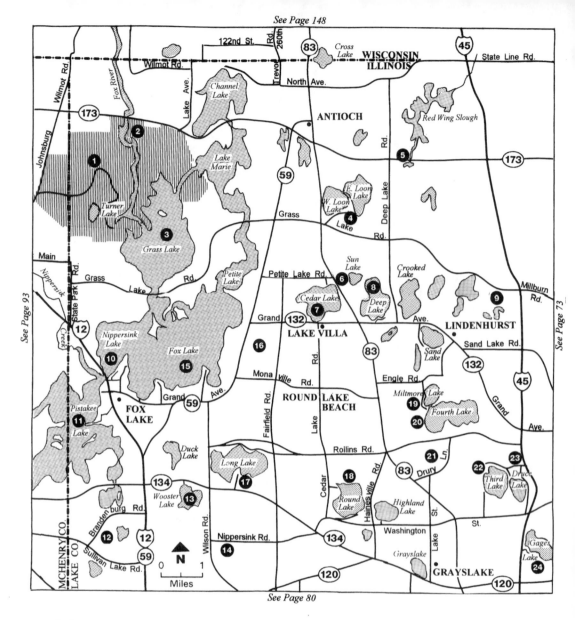

See Page 148

See Page 93

See Page 73

See Page 80

Map D Northwestern Lake County

1. Chain O'Lakes State Park *p. 90*
2. Oak Point Picnic Area *p. 92*
3. Grass Lake *p. 90*
4. East Loon and West Loon Lake *p. 87*
5. Red-Wing Slough *p. 83*
6. Sun Lake Forest Preserve *p. 87*
7. Cedar Lake *p. 87*
8. Deep Lake *p. 87*
9. McDonald Woods Forest Preserve *p. 88*
10. Nippersink Lake *p. 90*
11. Eagle Point on Pistakee Lake *p. 90*
12. Volo Bog State Natural Area *p. 89*
13. Wooster Lake *p. 86*
14. Nippersink Marsh *p. 83*
15. Fox Lake *p. 90*
16. Grant Woods Forest Preserve *p. 88*
17. Long Lake *p. 86*
18. Round Lake *p. 86*
19. Miltmore Lake *p. 86*
20. Fourth Lake *p. 86*
21. Drury Lane Pothole *p. 85*
22. Third Lake *p. 86*
23. Druce Lake *p. 85*
24. Gage's Lake *p. 85*

Trails—View the marsh from the highway. The area to the north is expected to have limited access in the future.

Owned By—Lake County Forest Preserve District (smaller area to south) and Illinois Department of Natural Resources (larger area to north)

Telephone—(847)367-6640

Drury Lane Pothole (D21)   Small, nondescript, and unprotected, the vernal pond off Drury Lane is consistently one of the best in the county. Tundra Swans, American Wigeons, Northern Pintails, and even a Cinnamon Teal are among the waterfowl that have rested on this tiny oasis. In early May, Greater and Lesser Yellowlegs, Willets, Semipalmated Plovers, Wilson's Phalaropes, and/or other sandpipers may be present.

Directions—From Rte. 45 turn west on Rollins Road to Drury Lane. (There are often flooded fields at the intersection that are worth checking.) Turn south on Drury to the high tension lines and wet area to the west.

Glacial Lakes

In addition to Lake Michigan, the crown jewel of them all, there are well over 400 glacial lakes in Illinois. A few drain into the Des Plaines River, but most are in the Fox River watershed. Almost all of them are in McHenry County and Lake County, with Lake County having the lion's share (it's name was not an accident). These lakes formed in morainal basins or when huge chunks of glacial ice became embedded in the ground and then melted. Several species of fish, including the very rare Pugnose Shiner, are locally restricted to some of the better quality lakes. For birders, these lovely water bodies are best visited in early spring, between the time when the ice leaves and the boats return. Birding in the fall is best after the boating season ends. It is also worth noting that hunting is allowed on many of the lakes. Winter birding is interesting if there are patches of open water.

W S S F

The bird that has become most closely associated with these lakes is the Common Loon. Not only can the species be seen here in large numbers during April, but the lakes provide the best opportunity to hear their haunting calls. Playing a tape will often coax a loon into vocalizing, or even swimming in for a closer look. Other water birds using these lakes include grebes, geese, Tundra Swans, ducks, coots, and gulls. Every body of water can produce interesting birds (a Northern Gannet once showed up at a detention pond in Vernon Hills); what follows is a small selection of the more heavily-birded lakes.

Gage's Lake (D24)   In some winters, Gage's Lake stays ice-free.

Directions—North of Rte. 120, turn east from Highway 45 on Sears Boulevard and park at Pebble Beach on the lake. Continue east to Rule Park and the Wildwood Park District offices.

Druce Lake (D23)

Directions—From Highway 45, turn west onto Sunshine and immediately south on Grant Avenue to see the lake from the east, or turn north from Washington Street onto Mainsail Drive and park at the boat launch.

Third Lake (D22)   With a maximum depth of 71 feet, Third Lake is the deepest lake in Lake County and among the last to freeze. In early winter, therefore, it often has a variety of waterfowl, especially concentrations of Canada Geese that can be huge. Linden Avenue is a good spot for Rusty Blackbirds in migration.

Directions—Turn west off Rte. 45 on Sunshine Avenue. Go past the Serbian monastery, turn right on Lake Avenue, and park at one of the boat launching points or at Third Lake Community building at the end. If you go the other direction on Lake Street, you will see Druce Lake. From the south, the lake can be viewed by turning north from Washington Street on Linden Avenue and parking at the Lake County Forest Preserve sign.

Trails—From the gate on Linden, walk across the grassy area to view the lake.

Owned By—Lake County Forest Preserve District (partially)

Telephone—(847)367-6640

Fourth Lake (D20)   Heavily bordered on one side by extensive marshes, Fourth Lake provides a breeding habitat lacking at most of the other lakes. In addition to other water birds, this is a good place to see Sandhill Cranes and Forster's Terns in May. The shrubs and trees can also be productive for migrating warblers.

Directions—Go east on Engle Road from Rte. 83. At North Nathan Hale Drive, turn south to South Nathan Hale, then east on Paradise Drive. Park at the end of Paradise Drive for a view of **Miltmore Lake** (D19) to the north and Fourth Lake to the south and east. This parking space faces southeast, so the best views are in the afternoon. For the Fourth Lake Fen, go west of Rte. 45 on Grand Avenue to Granada Boulevard. Go south on Granada one mile to the intersection of Genoa and Prairie View. Park on either side of Granada.

Owned By—Lake County Forest Preserve District

Telephone—(847)367-6640

Round Lake (D18)

Directions—Go north of Washington Street on Hainesville Road and turn west on Lake Avenue for a view of the lake from the east side. The south side of the lake can be seen by turning north from Washington Street. Turn sharply northeast on Clifton Drive to Bengson Park. On the west side of the lake, turn east from Cedar Lake Road onto South Channel Drive.

Trails—From Bengson Park, walk down to the beach.

Wooster Lake (D13)

Directions—Drive one mile east of Rte. 12 on Rte. 134. Turn south on Forest Drive and continue to the lake. Park along Wooster Lake Drive.

Long Lake (D17)

Directions—Turn north of Rte. 134 on Fairfield Road and turn west at the light at Lake Shore Drive. Park in pull-offs to view the lake. Continuing west, Lake Shore Drive merges with Rollins Road. A few blocks further, turn south on Wilson Road and stop north of the railroad tracks for a high vantage point. The south end of the lake can be viewed by driving west from Fairfield on Main Street just north of the railroad tracks. At Decorah, turn north and bear west on Muskego to Milwaukee Street. This vantage point is at the water's level.

Deep Lake (D8)

Directions—Turn north from Grand Avenue (Rte. 132) on Water's Edge Drive to Glacier Park.

Trails—Bird from the nearby condominium parking lot.

Cedar Lake (D7)

One of the most interesting of the small glacial lakes, Cedar Lake has a bog with a floating mat of sphagnum moss on its northwest shore and an island in the middle that hosts a small Great Blue Heron rookery. Some noteworthy history is associated with Cedar Lake. Locals used to collect cranberries from the bog, an activity that became less popular when a horse and a cow fell though the floating mat and drowned. And the island once housed certain female employees of Al Capone as they awaited more permanent assignments.

Directions—Drive south from Petite Lake Road on Belmoral Park and west on Liberty Avenue for a view from the north. On the south side of the lake, drive north from Grand Avenue on Cleveland Avenue to the end. On the east, drive south from Rte. 83 on Cedar Lane.

Trails—The east shore of the lake can be walked from Cedar Lane, but other locations on the shore are private property.

Note—No motors are allowed on this lake, but rowboats are available for rent. The marsh .5 mile south of Cedar Lake along Wind Dance Road east of Cedar Lake Road has had nesting Sandhill Cranes.

Sun Lake (D6)

Of the glacial lakes in Lake County, Sun Lake is one of the least disturbed by human activity. The woods, fields, and marshes making up the Forest Preserve that surrounds the lake can be good places to bird in spring and fall. The lake and surrounding vgetation have attracted rails and Forster's and Black Terns in summer.

Directions—Drive 4 miles west of Rte. 45 on Grass Lake Road and look for the entrance road to the preserve on the south side of the road. It is marked with a Lake County Forest Preserve sign, but this road may be closed to cars.

East Loon and West Loon Lake (D4)

Directions—Drive east of Rte. 83 on Grass Lake Road and turn north immediately on Villa Rica Road. Pull off onto the boat launch on the left for a view of West Loon Lake. To view both lakes, continue east to the intersection of Loon Lake Road and Lake Boulevard. Turn north onto the peninsula which affords a view of both lakes. Bird from the beach on West Loon Lake and respect private property.

Grant Woods Forest Preserve (D16)

W S S F

With 974 acres, this may be the most botanically diverse forest preserve in the Lake County system. There are oak woods, old fields, a high quality prairie, a cattail marsh, a stand of red pines, and a tamarack bog. A portion of the site has been dedicated as **Gavin Prairie** and Bog Nature Preserve. Warblers and other land birds use the woods in spring and fall. Summering species have included Least Bittern, Cooper's Hawk, Sandhill Crane, Alder Flycatcher, White-eyed Vireo, and Mourning Warbler.

Directions—There are two parking lots for the preserve. One is on Monaville Road just east of Rte. 59. Monaville Road bisects the preserve: the largest part (including the nature preserve and recreational areas) is south of the road. Another entrance is on Grand Avenue just east of Rte. 59. To park near the wooded southern part of the preserve, park on the north shore of Long Lake south of Rollins Road.

Trails—Trails circle through each section, and a trail connects the north and south sections. A total of six miles of trails meander through the preserve. The red pines are in the eastern part of the north section; a wooded ravine parallels the trail north of Monaville Road. The nature preserve, oak savanna, and cattail marsh are in the south section south of the sports field. The southern section can also be entered from Rollins Road.

Owned By—Lake County Forest Preserve District

Telephone—(847)367-6640

McDonald Woods (D9)

W S S F

Within its 296 acres, McDonald Woods contains two large ponds, extensive fields, a creek, scattered woods, and small stands of pines. Warblers and other migrant land birds can be found here in good numbers. Red-breasted Nuthatch (taking advantage of the conifers), Blue-winged Warbler, Pine Warbler, and Worm-eating Warbler are among the species that have been recorded. Although not as productive as they used to be, the ponds have had summering Pied-billed Grebe, Hooded Merganser, Common Moorhen, and other species. Field birds summering here have included Eastern Bluebird, Grasshopper Sparrow, Bobolink, and Eastern Meadowlark.

Directions—Turn west from Rte. 45 onto Grass Lake Road and park in the lot on the south side of the road approximately .5 mile west of Rte. 45.

Trails—Several trails loop around the area. The ponds are in the center, the fields are to the east, and the woods are scattered along the edges. Several neighboring houses have active bird feeders.

Owned By—Lake County Forest Preserve District

Telephone—(847)367-6640

Volo Bog State Natural Area (D12)

A common scenario in the formation of bogs is the gradual filling-in of a pond
that occurs as sedges and other vegetation form mats at the bog's edge. Over
time, the pond goes through several stages until it ultimately disappears. Volo
Bog is unique among Illinois's few bogs in that all stages are present: open water
in the center ringed by a floating mat ringed by tamarack trees ringed by shrubs
ringed by cattails. To buffer the bog, the state has protected the woods and fields
surrounding it. Rough-legged Hawks, Northern Shrikes, and crossbills are possible during the winter. Rusty Blackbirds visit the marsh in April and October,
while the bulk of the often abundant migrant land birds come through in May
and late August-September. The site boasts a diverse collection of breeding
species: Eastern Bluebird near the Tamarack View Trail; Bobolink in the fields;
Least Bittern, Wood Duck, Sandhill Crane, Sedge Wren, and Marsh Wren in the
marshes; and Alder Flycatcher, White-eyed Vireo, Veery, and Yellow Warbler in
the dense shrubs of the bog.

Directions—Turn west off of Highway 12 onto Brandenburg Road and park after
one mile in the lot for the visitors' center. The site is open from 8:00 A.M. to 4:00
P.M. Guided tours are given at 11:00 and 1:00 on Saturday and Sunday. The 120-
acre Pistakee Bog Nature Preserve is located .5 mile northwest of the parking lot.

Trails—Two trails are developed. The .5 mile Volo Bog Interpretive Trail passes
through the stages of the bog via a boardwalk, which is sometimes closed due to
high water. (Call the visitors' center if in doubt.) Be aware of the poison sumac that
grows along the boardwalk. The 2.75 mile Tamarack View Trail encircles the bog,
traversing various habitats and providing good views of several wetland areas.

Owned By—State of Illinois

Telephone—(815)344-1294

The Chain O'Lakes

As early as the 1880s, the Chain O'Lakes was a popular destination for visitors
attracted to its rolling hills and placid waters, gamefish, flocks of waterfowl,
and spectacular beds of flowering Lotus. To accommodate the growing tourism
industry, lake levels were raised in summer and lowered in winter by damming
the Fox River in 1914. Channels connecting the individual lakes were completed a year later. This in turn led to more recreational use and a larger permanent population. Filling of wetlands, increased development, declines in
water quality, and heavy boat traffic have all contributed to the degradation of
the lakes. The value of the Chain's natural qualities is well-recognized, however, and there are ongoing efforts by public and private agencies to help the
lakes recover.

Birding the nine shallow lakes that comprise the Chain is best in early spring
and late fall, periods of minimal boating activity. (During the winter, it is worth
examining unfrozen sections of lake.) Any spot where it is possible to view open
water is a good place to look. The water bird possibilities include Common
Loons, Pied-billed Grebes, Horned Grebes, geese, Tundra Swans, puddle ducks,
diving ducks, American Coots, gulls, and terns. Ospreys and Bald Eagles are

using the lakes with increased frequency. The most easily accessed marshes and uplands in the area lie within Chain O'Lakes State Park.

Fox Lake (D15)

Directions—Best viewing is from Leisure Point which separates Nippersink Lake on the west from Mineola Bay and Fox Lake on the east. Drive north from Grand Avenue on Forest Avenue and turn east on Lakeside Road before reaching Leisure Point and Korpan's Landing. Park at the end in the lot. Further east, turn north from Grand Avenue on Maple Avenue to view Mineola Bay. Further east, turn north on Tweed Road to view Fox Lake. Another spot to park is further east on Stanton Point. Turn north from Grand on Stanton Point Drive and continue to the end to view Fox Lake and Columbia Bay.

Nippersink Lake (D10)

Directions—From the intersection of Grand Avenue and Rte. 12 in Fox Lake, head east for a few blocks and turn north on Forest Avenue or go north on Rte. 12 and east on Oak Street to view the lake. A view can also be had from Nippersink Boulevard north of Oak Street.

Eagle Point on Pistakee Lake (D11)

This narrow promontory affords good views of Pistakee Lake, which is an excellent place to see Common Loons (up to 50 on one April day) and diving ducks. The peninsula across the lake to the north often has a Bald Eagle or two in March and harbors a Great Blue Heron rookery in spring and summer.

Directions—Turn west from Highway 12 at the stoplight for Eagle Point Road. Follow the many twists and turns on Eagle Point Road until you come to the end. Look north into Pistakee Lake or south into Meyers Bay. Take care not to block driveways.

Trails—Bird from your car. A scope is necessary here.

Grass Lake (D3)

Since the 1870s, Forster's Terns have almost always nested on a marshy island in this lake. Although in 1998 both the island and the terns were gone, history suggests they will return.

Directions—Turn north from Highway 12 onto State Park Road. After a mile, turn east on Grass Lake Road and proceed for 1.5 miles for a good viewing point for Grass Lake on the north side of the road.

Chain O'Lakes State Park (D1)

Straddling the Fox River, the state park contains 6,063 acres of marshes, lakes, fields, and woods. In winter, open areas are good for hawks and conifer stands have hosted Long-eared Owls. Ducks pause at Turner Lake during migration, particularly in the spring. Flycatchers, vireos, thrushes, warblers, and sparrows can be excellent in May and again in late August to September. Nesting species include Sandhill Crane, Common Snipe, Whip-poor-will, both cuckoos, Acadian Flycatcher, and Chestnut-sided Warbler.

Directions—Drive north from Rte. 12 or south from Rte. 173 on Wilmot Road and turn east at the entrance to the park. To bird the pines and small lakes, turn off the main road toward the park office and park in the first lot. For

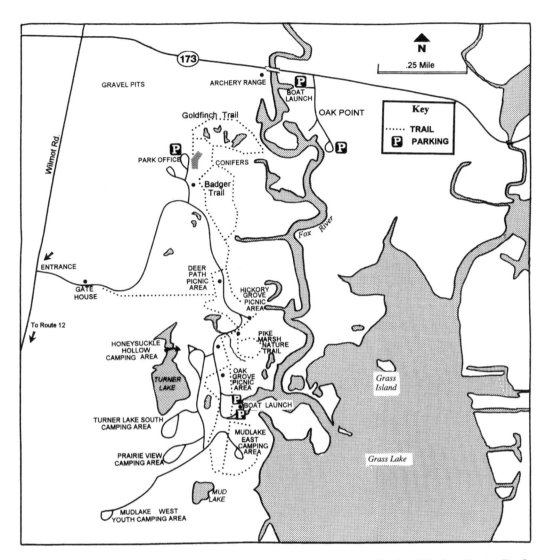

Chain O'Lakes State Park

marshes, Turner Lake, and the river, continue along the main drive to one of many parking areas.

Trails—The park has over 16 miles of trails. Near the park office, the Goldfinch Trail circles several small lakes, and the Badger Trail leads to a pine plantation. From the Deer Path Picnic area, a walk west into the marsh can be rewarding at dawn or dusk. From the Hickory Grove Picnic Area, a walk toward the east will take you to a channel in the marsh, a good spot to hear marsh birds. For marsh and land birds, walk east from the Pike Marsh North Picnic Area on the .25-mile trail. From Honeysuckle Hollow Campground, walk south along Turner Lake and bird the pond on the east side of the road leading to the Turner Lake South Camping Area. The upland woods along the roads and near the check station on the north side of the road can be good, too.

Note—The park is open from 6:30 A.M. to 9:00 P.M. Horse, bicycle, boat, and canoe rentals are available, and camping is a popular activity in season. The park is closed for hunting from the end of October to the end of December.

Owned By—Illinois Department of Natural Resources

Telephone—(847)587-5512

Oak Point Picnic Area (D2)   Located on the east side of Fox River, this unit of the Chain O'Lakes State Park with its wooded picnic area and boat launch is a fine place to be on an early May morning. Sandhill Cranes call frequently, and migrant land birds of many kinds sing from the tall trees. In recent years, Oak Point has been the best spot for nesting Cerulean Warblers in Lake County, but numbers are diminishing. Other breeding birds have included Acadian Flycatcher, Yellow-throated Vireo, and American Redstart. Forster's Terns are sometimes seen feeding over the river in spring and summer.

Directions—The entrance is on the south side of Rte. 173 just east of the Fox River. Continue all the way to the end and park.

Trails—Walk along the road and listen for bird activity.

McHenry County

Rush Creek Conservation Area (C1)

W S S F

These 390 acres of woods, evergreen plantings, fields, channelized creek, and pond provide refuge for migrating warblers and other passerines. Eastern Screech-Owls and Great Horned Owls are permanent residents.

Directions—Turn east onto McGuire Road from Highway 47, then turn south at the cemetery to enter the preserve. Park in the lot close to the equestrian trails.

Trails—The horse trail is better for birding than the hiking trails.

Owned By—McHenry County Conservation District

Telephone—(815)678-4431

Harrison-Benwell Conservation Area (C2)

W S S F

Although only 75 acres, this combination of fields, Bur Oaks, and a meandering stream has proven to be extremely productive for such migrant land birds as thrushes, warblers, and sparrows. The first section of the trail is particularly good.

Directions—From Highway 31, turn west on McCullom Lake Road and drive for about three miles to the site on the south side of the road. If coming from Rte. 120, turn north on Wonder Lake Road to McCullom Lake Road and turn east to the area.

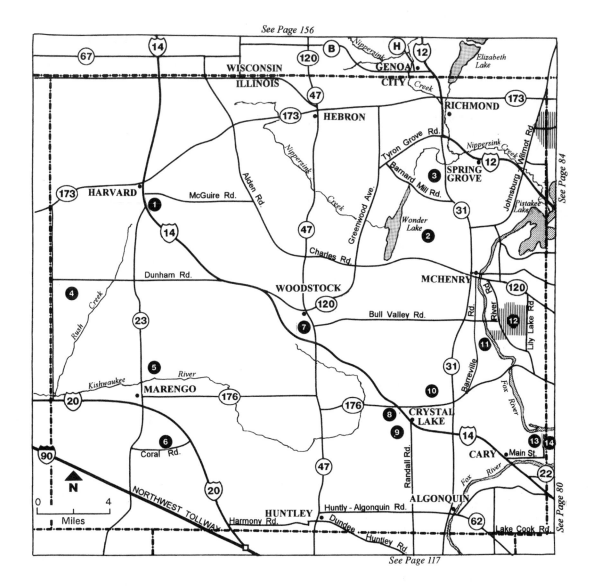

Map C **McHenry County**

1. Rush Creek Conservation Area *p. 92*
2. Harrison-Benwell Conservation Area *p. 92*
3. Glacial Park Conservation Area *p. 94*
4. McHenry County Sod Farms *p. 95*
5. Marengo Ridge *p. 95*
6. Coral Woods *p. 96*
7. Ryders Woods *p. 96*

8. Lippold Park *p. 97*
9. Crystal Lake *p. 97*
10. Sterne's Woods Nature Preserve *p. 97*
11. Stickney Run *p. 98*
12. Moraine Hills State Park *p. 98*
13. Hickory Grove Highlands *p. 99*
14. Lyons Prairie & Marsh *p. 100*

Trails—Cross the bridge over the stream and walk south along the western edge of the area. This trail will loop back to the east and north for a two-mile hike, but a shortcut trail will cut the length to about .25 mile.

Owned By—McHenry County Conservation District

Telephone—(815)678-4431

Glacial Park Conservation Area (C3)

W S S F

The 2,800-acre preserve is a showcase for the benefits that can accrue through careful ecological restoration. The McHenry County Conservation District has spent years re-creating, enlarging and/or improving the site's woodlands, grasslands, and wetlands. Their most ambitious project involves restoring Nippersink Creek by reconstructing the meanders that marked the stream prior to its channelization early this century. Birders will be most impressed by the creation of **Lost Valley Marsh**, formed when drainage tiles were removed from the fields along Harts Road. As a result of all this work, Glacial Park now ranks as the premier birding location in the county.

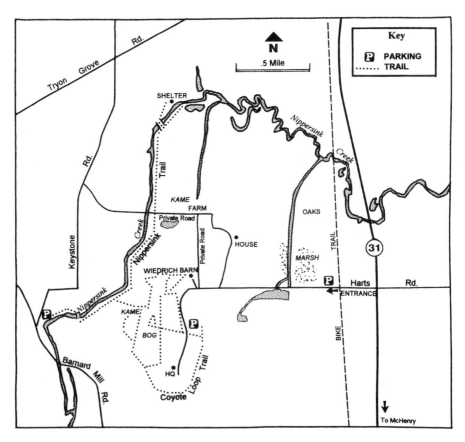

Glacial Park Conservation Area

Winter often produces Rough-legged Hawks, Short-eared Owls, and/or Northern Shrikes. During early spring and late fall, Sandhill Cranes pass over in large noisy flocks and waterfowl descend on the open waters and marshlands in droves. Greater White-fronted Geese, Snow Geese, and Tundra Swans sometimes accompany the Canada Geese. Clay-colored and other migrant sparrows gather in fields, and warblers are often common along the kames and in other wooded sections. Shorebirds, including such rarities as Long-billed Dowitcher and Wilson's Phalarope, probe Lost Valley Marsh. A tremendous variety of summering birds have discovered the diverse habitats that await them here, including Least Bittern, Sandhill Crane, Common Moorhen, Upland Sandpiper, Willow Flycatcher, Sedge Wren, Grasshopper Sparrow, Henslow's Sparrow, Dickcissel, Bobolink, and Orchard Oriole.

Directions—Drive west of Highway 31 on Harts Road. Park in the lot on the north side of the road just west of the Prairie bike trail or continue further west for views of Lost Valley Marsh. Turn north from Harts Road on the gravel road and park in the lot behind the farmhouse to view the marsh or continue west to the main parking lot and trails. For Nippersink Trail and Canoe Landing on the west side of the park, turn north from Barnard Mill Road onto Keystone Road and park in the lot immediately on the right.

Trails—From the Harts Road lot, walk north along the bike trail to view the Lost Valley Marsh and Grassland restoration area. Trails from the main lot lead through woodland, marsh, bog, and grassland and over a kame. The top of the kame affords extensive views of the area. A hike along Nippersink Creek is possible from the Keystone Road lot. Canoes can be lauched from this spot and taken out downstream at the Lyle C. Thomas Landing or at the Nippersink Canoe Base.

Owned By—McHenry County Conservation District

Telephone—(815)678-4431

McHenry County Sod Farms (C4)

W S S F

Beginning in May and then resuming in August, shorebirds stop here to feed in wet areas or where sod has been recently stripped. American Golden-Plovers and Buff-breasted Sandpipers are especially fond of sod farms, and they are a good possibility from the third week of August to mid-September. Other birds that make these sod farms interesting are nesting Western Meadowlarks and large flocks of American Pipits in October.

Directions—Drive along Dunham Road west of Rte. 23 to view the privately owned sod farms.

Marengo Ridge (C5)

W S S F

This upland area of oak woodlands, evergreens, and shrubs totals 240 acres and includes a picnic area and campground. Warblers and other land birds frequent the woods during spring and fall migration. An interesting suite of species summer here including Turkey Vulture, Ruby-throated Hummingbird, Veery, Blue-winged Warbler, and Ovenbird.

Directions—Park in the main parking lot on the east side of Rte. 23 north of Marengo.

Trails—For the nature and hiking trails, walk north on the 1.2-mile interpretive trail that begins near the water pump. Keep to the left and follow the trail as it circles back around to the parking lot. For a longer hike, take the 1.2-mile Kelly Hertel Trail extension north from the main trail. Another trail, a .6-mile loop, can be reached from the camping area to the east.

Owned By—McHenry County Conservation District

Telephone—(815)678-4431

Coral Woods (C6)

W S S F

Vireos, thrushes, warblers, and other passerines migrate through this 325-acre maple forest in large numbers during spring and fall. Eastern Wood-Pewees, Acadian Flycatchers, Great Crested Flycatchers, Red-eyed Vireos, and Scarlet Tanagers have stayed to nest. In recent years, Swainson's Hawks have summered in the vicinity.

Directions—From Rte. 20, turn west on Coral West Road. In .5 mile, look for Somerset Road on the north and follow it to the end of the road to the preserve. The preserve is open from 8:00 A.M. to sunset.

Trails—The trails begin near the restroom. Head north and look for the 1.2-mile nature trail which branches off to the left and passes through an open field before entering the woods. Most of the nature trail is in the woods where there is a good understory of spring wildflowers and several small streamlets. One footbridge has been erected, but the other crossings are on rocks which may be difficult to navigate in a wet season.

Owned By—McHenry County Conservation District

Telephone—(815)678-4431

Ryders Woods (C7)

W S S F

This is another one of those tiny places that acts as a magnet for migrant passerines. Located in Woodstock, the 25 acres contain woods, a small pond, field and marsh.

Directions—From Business Rte. 14 (Lake Street), turn west on Kimball. Proceed .2 mile and park in the lot on the north side of the street. The McHenry County Chapter of the Illinois Audubon Society has monthly walks here.

Trails—Paths crisscross the area, and benches are strategically placed along the main path, including one overlooking the small pond.

Lippold Park (C8)

During migration, Common Loons, Redheads, mergansers, Caspian Terns, Forster's Terns, and other water birds use the two ponds and 300 acres of wetland that are protected here. During the summer, the marshes attract foraging Black-crowned Night-Herons and nesting Soras and Yellow-headed Blackbirds.

Directions—Drive .3 mile west of Rte. 14 on Rte. 176, turn south into the park and park in the west lot. From this lot, you will have to walk to see the wetlands, but they may be seen from another parking lot. Go further west on Rte. 176 and turn south into the lot for the golf driving range where you can use a scope to see the ponds.

Trails—From the main parking area, walk west toward the gazebo until you see the water.

Owned By—Crystal Lake Park District

Telephone—(815)459-0680

Crystal Lake (C9)

Before and after the heavy boating activity of summer, this deep lake can be a good spot for waterfowl, including Common Loons, Horned Grebes, and mergansers.

Directions—From Rte. 14, turn southwest on Dole Street and continue to the public beach. Park in the lot on the left and cross the road to the lake. For a different view, continue west on Lake Avenue and North Avenue to reach the west beach parking area.

Sterne's Woods Nature Preserve (C10)

A hilly preserve of upland oak and hickory woods with a few pines and cedars and low, grassy areas, it is best for warblers and other woodland migrants. Broad-winged Hawks have been observed in summer.

Directions—From Rte. 31, drive west on Rte. 176 for .5 mile to Terra Cotta Road. Turn north to Hillside Road and turn west. Look for the entrance road in .5 mile on the south side of the road.

Trails—Two miles of trails circle the preserve and connect to **Veteran Acres Park** off of Rte. 176 to the south, with recreational fields, a pond, open grassy areas, pine plantings, and deciduous woods. The pines may attract owls in winter. Bluebird houses have been erected and are maintained in the grassy area in the eastern part of the park.

Note—The preserves are not open until 9:00 A.M. and close at dusk.

Owned By—Crystal Lake Park District

Telephone—(815)459-0680

Stickney Run (C11)

W S S F

A wooded ridge on the Fox River and marshy lowlands distinguish this 310-acre site. The woods attract numerous migrants in the spring and fall. Summer residents include Least Bitterns, Sandhill Cranes, Red-headed Woodpeckers (look for them in the dead trees off of State Park Road east of the park entrance), and Yellow-headed Blackbirds.

Directions—Turn east from Rte. 31 on Bull Valley Road. Before crossing the river, turn south on Barreville Road. In one mile, turn east on State Park Road and look for the entrance to the preserve on the south side of the road.

Trails—One trail heads southwest from the parking lot and skirts the small fishing pond. Another trail goes south and has a boardwalk into the wetland area.

Owned By—McHenry County Conservation District

Telephone—(815)678-4431

Moraine Hills State Park (C12)

W S S F

This 1,690-acre park is a mosaic of wooded moraines, bogs, open fields, lakes, riverfront, evergreen plantings, and marshes. (A large chunk has been dedicated as a state nature preserve.) On certain days in early spring and late fall, the intermittant bugling of Sandhill Cranes lasts for hours as hundreds of birds pass overhead. Less conspicuous, but of equal interest to the birder, are the many land birds that utilize the woodlands during their migration periods.

The state park is divided by River Road into two sections. The eastern portion is the largest and includes 48-acre Lake Defiance which is excellent for Common Loons, Horned Grebes, numerous ducks, and Ospreys. If water levels are low enough, shorebirds can also be seen in the vicinity of the lake, as well as the ponds in the northwest part of the park. Marshes surrounding the lake have their share of breeding species including Pied-billed Grebes, Least Bitterns, and Virginia Rails.

The smaller western portion abuts the Fox River and is the site of **McHenry Dam** and **Black Tern Marsh**. Below the dam, the river opens earlier and freezes later than other stretches. At such times various ducks are often present. A large tract of evergreens usually holds large numbers of Red-breasted Nuthatches from late fall through early spring. Warblers and other migrant passerines are excellent in the woods along the Fox River and shorebirds are possible around the ponds if water conditions allow. The marshes and riparian woods also provide habitat for an exciting mix of breeding species including Least Bittern, Common Moorhen, Sandhill Crane, Black Tern, and Yellow-headed Blackbird. This is perhaps the most reliable spot in the northern part of the Chicago region for nesting Prothonotary Warblers.

Directions—The main entrance to the park is on the east side of River Road two miles north of Highway 176 and three miles south of Highway 120. Park at the park office to get a map. To reach McHenry Dam, continue one mile north on River Road and turn west onto McHenry Dam Road. Park in designated areas only.

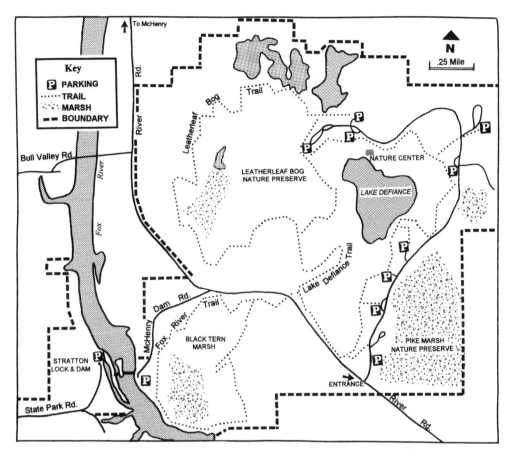

Moraine Hills State Park

Trails—The interpretive center at park headquarters overlooks Lake Defiance and offers various displays and maps of the park. There are 10 miles of trails in the park. Two trails in particular are recommended. In the eastern portion, the 3.5-mile trail to the Northern Lakes and Leatherleaf Bog traverses woods, fields, marshes, and ponds. The 2-mile Fox River Trail connects the eastern part of the park to the Fox River and McHenry Dam. From the dam parking area in the western section, follow the trail from the south end of the parking lot to the viewing platform at Black Tern Marsh and continue to the second viewing platform on the opposite side of the marsh.

Owned By—Illinois Department of Natural Resources

Telephone—(815)385-1624

Hickory Grove Highlands (C13)

A marshy backwater of the Fox River and a section of the Cary Moraine covered by woods and fields are combined in this 225-acre preserve. During migration, the river attracts waterfowl, and the wooded upland provides habitat for many

W S S F

passerines including Connecticut Warbler and Yellow-breasted Chat. The open area is a good place to see displaying Woodcock and nesting Bobolinks.

Directions—From Cary, go east on Main Street for two miles to Hickory Nut Grove Road (the Lake County Line) and turn north. Pass the lot for **Lyons Prairie & Marsh** and turn west where the road turns into Hickory Grove Road. The parking lot for **Hickory Grove Riverfront** is on the north side of the road one-half mile west of the turn. Continue .25 mile further to the entrance to Hickory Grove Highlands on the south side of the road.

Trails—Hiking trails connect all three sites, and an equestrian trail goes around the western and southern perimeter of Hickory Grove and Lyons Prairie and Marsh.

Owned By—McHenry County Conservation District

Telephone—(815)678-4431

Lyons Prairie & Marsh (C14)

W S S F

Although this 360-acre site is in Lake County, it lies on the west side of the Fox River and is owned by the McHenry County Conservation District. Two hundred and forty-five acres of its wet prairie, marsh, floodplain woods, backwater ponds, and Fox River frontage are dedicated as a nature preserve. The woods are good for migrant land birds, and the marshes have breeding Virginia Rails, Sedge Wrens, and other wetland species. Cooper's Hawks and Blue-gray Gnatcatchers nest in the woodlands. Herons use a rookery on the river and may be seen foraging in the preserve.

Directions—From Cary, go east on Main Street for two miles to Hickory Nut Grove Road (the Lake County Line) and turn north. The parking lot for Lyons Prairie and Marsh will be on the east side of the road in one mile. For a view of the rookery, cross over to the east side of the Fox River on Rte. 176 and turn

Map L **DuPage County** (facing page)

1. Pratt's Wayne Woods Forest Preserve *p. 102*
2. Spring Brook Nature Center *p. 220*
3. Songbird Slough Forest Preserve *p. 102*
4. Fullerton Park Forest Preserve *p. 104*
5. Fermilab *p. 103*
6. West Chicago Prairie Forest Preserve *p. 105*
7. Timber Ridge Forest Preserve *p. 106*
8. West DuPage Woods Forest Preserve *p. 106*
9. Elsen's Hill *p. 106*
10. McKee Marsh *p. 107*
11. Mt. Hoy *p. 107*
12. Herrick Lake Forest Preserve *p. 108*
13. Danada Forest Preserve *p. 109*
14. Morton Arboretum *p. 109*
15. Hidden Lake Forest Preserve *p. 111*
16. Maple Grove Forest Preserve *p. 111*
17. Lyman Woods Forest Preserve *p. 112*
18. Fullersburg Woods Forest Preserve *p. 112*
19. Eola Road Marsh *p. 113*
20. McDowell Grove Forest Preserve *p. 113*
21. Springbrook Prairie Forest Preserve *p. 113*
22. Greene Valley Forest Preserve *p. 114*
23. Waterfall Glen Forest Preserve *p. 115*

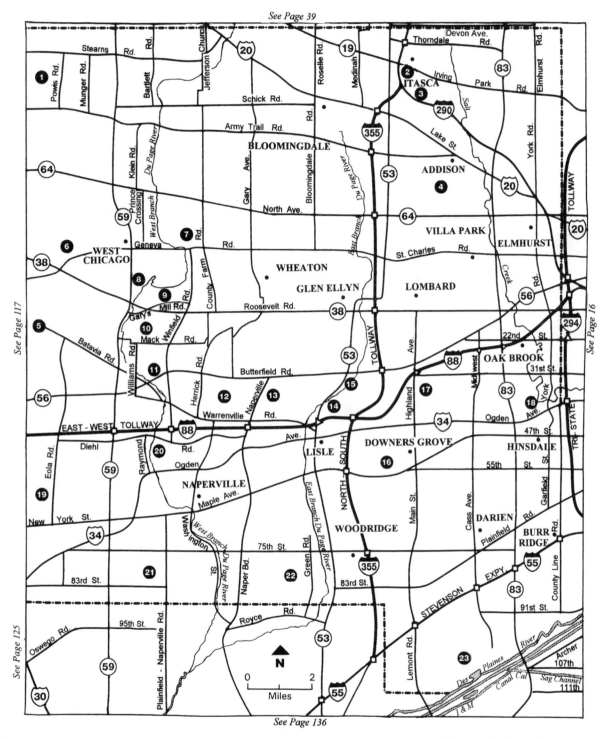

See Page 39

See Page 136

Map L **DuPage County**

south on Roberts Road. Enter the **Fox River Forest Preserve** and look across the river for the herons' nests.

Owned By—McHenry County Conservation District

Telephone—(815)678-4431

DuPage County

Pratt's Wayne Woods (L1)

W S S F

This is the largest holding within the DuPage County Forest Preserve District, with 3,385 acres. Pratt's Wayne contains fields, small lakes, a creek, open woods, and a marsh complex along **Brewster Creek** and the Illinois Prairie Path that is being expanded to over 700 acres. The marshes are used by waterfowl in spring and fall. Shorebirds and herons are seen from spring through fall, and Sandhill Cranes, Black Terns, Warbling Vireos, and Yellow-headed Blackbirds have nested. Migrating passerines prefer the woods along the bridle path south of the main parking lot and the trees by the small ponds. In most years, Northern Harriers and Short-eared Owls linger in the fields through the fall into early winter.

Directions—Drive north from Army Trail Road on Powis Road and enter the preserve on the west side of Powis. Park in the farthest lot for access to the lake, woods, and marsh. To view the fields, go through the culvert under the railroad tracks on the east side of Powis Road to the dog-training area. The fields on the east side of the preserve can also be seen from Munger Road, and wintering Short-eared Owls can often be seen at dusk. For the closest access to the marsh, park along Dunham Road at the Prairie Path just south of Little Woods Cemetery. Another access to the Prairie Path is from Army Trail Road east of Powis. Note that parking lots do not open until one hour after sunrise.

Trails—The trail from the main lot around the youth campground area and around the ponds is good for warblers and other passerines. From Catfish Pond, a trail heads west to the Illinois Prairie Path and Brewster Creek and Marsh. (Be watchful for fast-moving bicycles along the prairie path.) The path can also be accessed from Dunham Road or Army Trail Road. The marsh is approximately .5 mile south of Dunham Road and 1.5 miles north of where the Prairie path intersects with Army Trail Road. The fields can be birded from the roadsides.

Owned By—DuPage County Forest Preserve District

Telephone—(630)933-7200

Songbird Slough Forest Preserve (L3)

W S S F

The lake and wetlands within this 383-acre preserve draw migrant waterfowl and, if water levels allow, shorebirds. A varied passel of species has summered including Pied-billed Grebes, Black Terns, Grasshopper Sparrows, Dickcissels, and Yellow-headed Blackbirds.

Directions—From Irving Park Road, turn south on Parkside Road/Mill Road. Follow the entrance road to the end for a view of the lake.

Owned By—DuPage County Forest Preserve District

Telephone—(630)933-7200

Fullerton Park (L4)

This forest preserve has a cattail marsh surrounded by industrial parks, railroad tracks, and a model helicopter field. The marsh is frequented by Great Blue Herons, Great Egrets, Black-crowned Night-Herons, and other marsh birds, depending on water levels.

W S S F

Directions—The official lot is on Grace Avenue, but the best views of the marsh are from the west. From Fullerton, turn south at the light on Lombard and then turn east on Jeffrey. The first drive to the south leads to village property and a good view of the marsh, but this drive is often locked. Continue east and pull into the parking lot behind the commercial building at 1413 Jeffrey to view the marsh. For access from the south, turn north from North Avenue at the light at Main Street and go as far north as possible and park in one of the parking lots for the businesses located there. By walking over the railroad tracks, you will have another view of the marsh.

Note—Much of the property is subject to development.

Owned By—DuPage County Forest Preserve District

Telephone—(630)933-7200

Fermilab (L5)

Fermilab is the abbreviated name of the Fermi National Accelerator Laboratory, which produces the highest energy particle beams in the world with its Tevatron Accelerator. The campus includes numerous buildings, of which Lederman Education Center and Wilson Hall are open to the public and worth visiting. But birders come for the 6,800 acres of varied habitat that surround the buildings. The grasslands and water area inside the accelerator ring are off limits, although birders are allowed to visit most other areas with permission. (Ask the guard for a pass.)

W S S F

This is one of the region's best locations for winter birding away from Lake Michigan. Long-eared Owls often roost in the conifers; Horned Larks, Lapland Longspurs, Snow Buntings, and Brewer's Blackbirds forage near the bison feeders; Northern Harriers, Rough-legged Hawks, and Northern Shrikes work the fields near Casey's Pond and Lake Law; and seed-eaters frequent the feeders at the Education Center. In early spring and late fall, huge flocks of Canada Geese settle at Fermilab, often grazing in the bison field. Careful searching of the goose hordes has turned up Greater White-fronted, Snow, and Ross's Geese. During the same periods, Common Loons, grebes, and a wide variety of ducks use Lake Law, A.E. Sea, the Sea of Evanescence, Casey's Pond, and other lakes. The edges of Lake Law, Swenson Pond, and transient ponds are good places for shorebirds and pipits in late spring, late summer, and fall. Migrant passerines are especially

Fermilab

common at the Sparrow Hedge south of Lake Law, Big Woods, and Horseshoe Pond. Nesting species include Sedge Wren, Grasshopper Sparrow, Dickcissel, and Bobolink in grasslands and Bell's Vireo and Yellow-breasted Chat in thick scrub.

Directions—The eastern entrance is on Batavia Road west of Rte. 59; the western entrance is on Pine Street east of Kirk Road. Both entrances are gated, and cars must exit through the gate entered. The following areas are of interest to birders proceeding west from the eastern entrance on Batavia Road: A.E. Sea can be reached by parking at the Farm Machinery Display on Batavia Road. For the Dusaf Pond, drive .25 mile west to Blackhawk and turn north for .3 mile. The elevated pond is on the east side of the road. For Lake Law, Sea of Evanescence, Village Pines, and Sparrow Hedge, park near the Red Barn on Sauk Boulevard south of Batavia Road. For the Swenson Road pond and the Main Ring Lake, turn south from Batavia Road onto Eola Road and follow it to Swenson Road. The small pond at the intersection (the Swenson Road Pond) may harbor shorebirds in migration. Continue straight ahead to the grasslands along Eola Road or turn west on Swenson and park in the lot adjacent to the main accelerator ring. The bison pen can be viewed from Batavia Road west of Eola Road. Casey's Pond is one mile north of the intersection of Batavia and Eola Road at Wilson Street. It is the lake with the spiral-shaped concrete pump house. For Swan Lake, head for 15-story Wilson Hall and park in the lot on the north side near the lake. The education center is just west of Swan Lake and can be reached by walking a short trail west through the woods. The oak woodlands (Big Woods) are immediately north of Swan Lake. Horseshoe Pond on the northern fringes of the woods can be reached by parking in the lot just north of the fire station on Receiving Road.

Trails—Much of the property can be birded from the roadside, but the following walks are necessary to fully explore the birds. The brushy area under the electrical towers on the east side of Fermilab can be walked from the Farm Machinery Display; the Sparrow Hedge and the cattails at the southern end of Lake Law can be reached by following the berm around Lake Law south of the Red Barn. By veering east on this trail, you will reach the Sea of Evanescence. To bird a productive grassland trail, go .25 mile south of Swenson Road on Eola Road, park just before the "Curve" road sign and walk through the dense grasses. To see the Main Ring Lake, walk up the stairs from the parking lot on Swenson. The Illinois Prairie Path cuts diagonally across the southwest portion of the grounds, and a good spot for shorebirds in the spring can be seen from the Prairie Path north of Butterfield Road (Rte. 56).

Owned By—U.S. Department of Energy

Telephone—(630)840-3351

West Chicago Prairie (L6)

W S S F

Because of their great agricultural value, very few black soil prairies have survived. The largest high quality example in this region is West Chicago prairie. Although Goose Lake Prairie in Grundy County is larger, more of it has been disturbed. For each extant native prairie there is at least one specific reason why it was never developed. In this instance, the prairie was railroad right-of-way, one tiny fragment of the vast lands that the federal government ceded to the railroads during the last century.

In March, the prairie is a good place to find newly arrived Wood Ducks, Killdeer, Fox Sparrows, Eastern Meadowlarks, and Rusty Blackbirds. April brings rails, Common Snipe, American Woodcock, and a good assortment of sparrows. And by mid-May, the spring land bird migration is at its peak, with cuckoos, flycatchers, vireos, wrens, thrushes, and warblers present in substantial numbers. Some stay to breed, including both cuckoos, White-eyed Vireos, Warbling Vireos, Sedge Wrens, Marsh Wrens, Wood Thrushes, and Yellow-breasted Chats. Many of the same passerines come through again in the fall, along with migrating hawks.

Directions—Turn west from Rte. 59 onto Hawthorne Lane; drive two miles and turn south on Industrial Drive. Park in the lot on the east side of the road. Another access is off of Hawthorne at the south end of McQueen Drive.

Trails—Walk north from the parking lot through the quarry to reach the trails or walk south and west from the Illinois Prairie Path on the northern border into the preserve.

Note—The **DuPage Airport** north of Hawthorne Lane and west of Powis Road may have Lapland Longspurs and Snow Buntings in the winter and nesting Dickcissels and Western Meadowlarks.

Owned By—DuPage County Forest Preserve District

Telephone—(630)933-7200

Timber Ridge Forest Preserve (L7)

W S S F

Among the more noteworthy migrants that have been seen in this 611-acre preserve along the West Branch of the DuPage River are Cerulean, Prothonotary, and Worm-eating Warblers, Louisiana Waterthrushes, and Kentucky, Connecticut, and Hooded Warblers. Orchard Orioles can often be heard or seen near the bridge over Kline Creek. White-eyed Vireos, American Redstarts, and Yellow-breasted Chats have nested here. Gulls are seen in winter on the preserve's large pond.

Directions—Park in the lot on the west side of County Farm Road north of Geneva Road.

Trails—The Prairie Path traverses these woods. The path gets crowded on weekends, thus birding is best very early in the morning.

Owned By—DuPage County Forest Preserve District

Telephone—(630)933-7200

West DuPage Woods (L8)

W S S F

This preserve lies along both sides of the West Branch of the DuPage River. There are 460 acres on the west bank, some of which are covered by recreational fields and maintenance buildings. During migration, good birding is available along the hedge near the maintenance building. The more heavily birded section is on the east side of the river. Called **Elsen's Hill** (L9), this 57-acre parcel has ponds, fields, woods, and shrubby areas. It is one of the best spots in DuPage County to see warblers and other woodland migrants. In May, thrushes and warblers like the upper areas and the blooming hawthorn trees. Look (and listen) for Connecticut Warblers in the shrubby areas in late May. Red-headed Woodpeckers and Yellow-throated Vireos nest in the big trees of the lower areas. In the fall, vireos and many Yellow-rumped Warblers can be found eating Poison Ivy berries. Ospreys, Sharp-shinned Hawks, Cooper's Hawks, Red-shouldered Hawks, and Broad-winged Hawks are regularly seen migrating overhead.

Directions—To reach the western portion of the preserve, turn east into the preserve from Rte. 59 north of Rte. 38. To reach Elsen's Hill, turn onto Gary's Mill Road from Rte. 38 or from Winfield Road east of Rte. 59 (neither intersection has a traffic light). Park in the lot on the north side of Gary's Mill Road approximately .5 mile north and west of Rte. 38 and .5 mile east of Winfield Road.

Trails—Four miles of trails are available at Elsen's Hill; a trail map is posted at the north end of the parking lot. Begin birding in the parking lot. Check the edges for warblers and the sky overhead for raptors in migration. This upper area has two small ponds—one near the parking lot and a larger pond closer to Gary's Mill Road west of the entrance. From the first pond, take one of the trails to the southwest toward the second pond. Walk around this pond on the trail and onto the old road. Follow the road north to the open/scrubby area. This area is often very productive for birds. Several trails pass through this area of shrubs and trees and connect to the parking lot. The trail north of the parking lot leads to Elsen's Hill which was once used for tobogganing. Continuing north, the trail descends the hill, goes west along the West Branch of the DuPage River, circles around a meadow, and returns to the parking lot through the area of shrubs and hawthorn trees.

Note—The trails can be quite muddy and pitted by horse hooves.

Owned By—DuPage County Forest Preserve District

Telephone—(630)933-7200

Blackwell Forest Preserve (L10 & L11)

W S S F

This 1,314-acre preserve contains two of the best birding areas in the county. **McKee Marsh** (L10) has a pond and cattail marsh surrounded by open fields and woods. In some years, the water dries up entirely or becomes an area of mud-flats. Depending on water levels, herons, ducks, and/or shorebirds are present in season. Nesting species include Virginia Rail, Sora (the rails nest near the spillway), Willow Flycatcher, Eastern Bluebird, Savannah Sparrow, and Bobolink. In dry years, the area is attractive to Sharp-tailed Sparrows from mid-September to early October. They prefer the islands in the middle of the marsh. The wooded edges to the west of the marsh are good for warblers in migration.

Mt. Hoy (L11) is a 150-foot man-made hill that has become one of the region's three premier hawk-watching stations. During the fall, dedicated observers have manned the post on and off since 1985. If raptors are in the air, they are easy to see from Mt. Hoy. The great majority of individuals recorded consists of three species: Sharp-shinned Hawks and Broad-winged Hawks peak during the period from late September to early October, while Red-tailed Hawks are most abundant during the last two weeks of October and first week of November. Interspersed among the more common hawks are Turkey Vultures, Ospreys, Bald Eagles, Northern Harriers, Cooper's Hawks, Rough-legged Hawks, and Merlins.

Directions—Exit I-88 at Winfield Road northbound. To reach McKee Marsh, turn west in approximately three miles onto Mack Road. Park in the lot on the north side of Mack Road 1.5 miles west of Winfield Road (.3 mile east of Rte. 59). To reach Mt. Hoy, enter the preserve from Butterfield Road (Rte. 56). There are two parking areas. Park in the lot for the boat launch or in the next lot to the north.

Trails—To bird McKee Marsh, walk north on the path from the parking lot to the 1.4-mile Bob-o-link trail. This trail can be walked in either direction; on the east side, there is a peninsula that juts out into the water. To the west is the spillway area, barn, and woods. The trail does not follow the eastern edge of the marsh; some off-trail hiking is required if you want to stay close to the marsh and have a view from the north. The Catbird Trail parallels the West Branch of the Du-Page River and can be reached off the west side of Bob-o-link Trail.

To bird Mt. Hoy, cross the entrance road from the boat launch lot and walk up the blacktopped path to the top of Mt. Hoy. (The path is not easy to see.) Or walk up the shorter, steeper slope (the actual toboggan run) near the second parking lot. Be sure the gates are open at the top of the hill before making the climb.

Owned By—DuPage County Forest Preserve District

Telephone—(630)933-7200

Herrick Lake Forest Preserve (L12)

W S S F

Besides good numbers of migrants, an interesting mix of nesters use the fields, lakes, marshes, and open woods of this 767-acre preserve. In addition to Herrick Lake, another small body of water known as the Hesterman Drain Project is located approximately 1.5 miles south and east of the Herrick Road parking lot. Whereas Herrick Lake is used primarily for fishing and canoeing, the Hesterman Drain Project is good for birding. Surrounded by marsh and dead trees, it attracts various herons and Common Moorhens throughout the summer and is a good spot for Nelson's Sharp-tailed Sparrows in the fall. Other species that summer on the preserve include Sedge Wren, Blue-winged Warbler, Bobolink (grassy area west of the Hesterman Drain Project), Savannah Sparrow, Grasshopper Sparrow, and Henslow's Sparrow (grassland on the west side of preserve).

Directions—Exit I-88 at Naperville Road northbound. Turn west at the first intersection (Warrenville Road) and continue to Herrick Road. Turn north and park in the lot off Herrick Road one mile north of Warrenville Road for the western area or continue to Butterfield Road and turn right to the main entrance. Parking is also available on the south side of the preserve along the road at Warrenville Road and Washington Street or east of the preserve in Danada Forest Preserve. Parking lots do not open until one hour after sunrise.

Trails—Five-and-a-half miles of trails form loops in the preserve. A map of the trails is posted in both parking lots for the preserve. From Herrick Lake, walk south to the main trail, identified on the map as the Regional Trail. This trail forms a loop and connects to two more trails which also form loops. The second loop circles through open fields to the east and west, deciduous woods to the south, and the Hesterman Drain Project. This trail can be accessed from Warrenville and Washington. The third loop is to the north and east of the second trail. It goes around a grassy area and connects with a trail to the Danada Forest Preserve

Owned By—DuPage County Forest Preserve District

Telephone—(630)933-7200

Danada Forest Preserve (L13)

This 763-acre preserve contains open oak woods, the **Rice Lake** retention pond, restored prairie, brushy fields, and a small wetland. Great Horned Owls nest in the woods east of the barn, and American Woodcock display in open areas from March through April. Grebes, cormorants, herons, geese, ducks, shorebirds, gulls, and terns can be seen in or around Rice Lake during spring and fall migration. Cuckoos, thrushes, warblers, and sparrows migrate through the woods and along the fence rows, particularly in spring. Bobolinks perch on the fences around the barn areas in spring, and Grasshopper Sparrows and Dickcissels nest in the fields north of the Danada mansion.

Directions—Exit I-88 at Naperville Road northbound and enter the preserve in 1.5 miles on east side of the road (.5 mile south of Butterfield Road). The parking lot does not open until one hour after sunrise. Do not disturb the activities at Danada Mansion where private receptions and meetings are held.

Trails—The trails north and east of the horse barns circle through an oak savanna with native wildflowers; the trail heading north passes through a brushy and grassy area on its way to Rice Lake. The trail which passes under Naperville Road to the west leads to a 35-acre restored prairie and wetland area and connects to the trail system at Herrick Lake Forest Preserve.

Owned By—DuPage County Forest Preserve District.

Telephone—(630)933-7200

Morton Arboretum (L14)

Straddling the East Branch of the DuPage River, the Morton Arboretum is a 1,500-acre living museum devoted to woody vegetation. Over 3,600 varieties of trees, shrubs, and vines from across the globe are on display. There are also lakes, tracts of native woods, and restored prairie. The abundance of conifers and other seed-bearing plants makes the arboretum an outstanding place for winter birding. Evergreens and alders should be checked for owls, woodpeckers, and winter finches (Purple Finches, Red and White-winged Crossbills, Pine Siskins, and Evening Grosbeaks). There are alder plantings along the DuPage River on both sides of Rte. 53. White-winged Crossbills are partial to hemlocks, which can be found near the Hedge Garden and on Hemlock Hill. Cedar Waxwings (joined on rare occasions by Bohemian Waxwings) and American Robins flock to crabapples and other fruit-bearing trees. During those winters when Arbor Lake remains unfrozen, Green-winged Teal, Common Goldeneye, and other ducks are often present. In early spring and late fall, the lakes attract migrating waterfowl. To see migrant land birds in the spring and fall, walk the Illinois Trees Trail on the east side and the entire west side (especially the Joy Path and areas east of the entrance road as it approaches the Thornhill building). Breeding species are limited, but three vireos (Yellow-throated, Warbling, and Red-eyed) and Scarlet Tanagers summer in the native woods.

Directions—Exit I-88 onto Rte. 53 northbound and look for the entrance to the

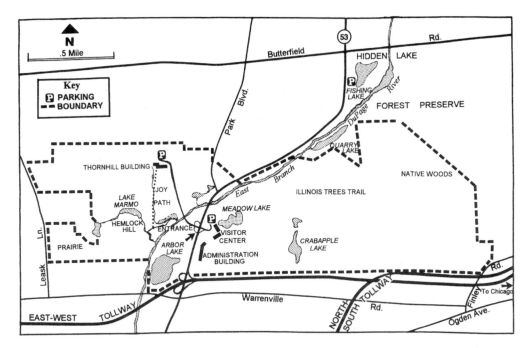

Morton Arboretum

Morton Arboretum on the right (east). Route 53 bisects the arboretum; the main parking lot, visitors' center, restaurant, library, gift shop, and theater are all on the east side near the entrance. A one-way, 12-mile road meanders through the Arboretum and crosses under Rte. 53 to the west side. Parking is allowed in lots or pull-offs along the way, and the lots are numbered (P1, P2, etc.). Good birding spots on the east side include the Hedge Garden (near P1), native Sugar Maple and oak woods at the far eastern end, the crabapple collection (near P3), and the evergreens in the Japanese collection (P16) and Spruce Plot (P11). Good birding spots on the west side include Lake Marmo (P20A), Pine Hill, Hemlock Hill, and Spruce Hill (P22–24), Bob-o-link Meadow (P24), Arbor Lake (P26), and the DuPage River (P26).

Trails—Twenty-five miles of trails exist here. Most of the trails, including the 2.9-mile Illinois Trees Trail with spring-flowering shrubs, meadows, marshes, and Meadow Lake are on the east side. On the west side of the arboretum, the Evergreen Trail passes through Pine Hill, Hemlock Hill, and Spruce Hill, and Joy Path begins south of the Thornhill Education Center and passes through open woodland alongside a small brook.

Note—The arboretum opens at 7:00 A.M. Non-members must pay an entry fee ($7.00/car in 1999), but cars with membership stickers for the Chicago Botanic Garden are admitted free. Maps are available from the attendant at the gate. A tour bus operates on a seasonal schedule and is accessible for wheelchairs. Floyd Swink's Finding List of the Birds of Morton Arboretum is sold in the gift shop, and a seasonal slide show in the nearby theater provides a good introduction to the Arboretum.

Telephone—(630)968-0074

Hidden Lake (L15)

Situated along the East Branch of the DuPage River just north of the Morton Arboretum, Hidden Lake is a 390-acre Forest Preserve encompassing two lakes, marshes, fields, and forest. If the river is not frozen, the preserve is a good place to see wintering waterfowl, including Horned Grebe, seen on at least one occasion. More waterfowl appear on the lakes or river during migration. American Woodcock display here in early spring, and migrant land birds are common in the woods and fields. Nesting species have included Spotted Sandpiper, Marsh Wren, and Bobolink. In late summer and early fall, southbound shorebirds often put in an appearance. Among the species noted have been White-rumped Sandpiper, Dunlin, and Wilson's Phalarope.

Directions—Park in the lot on the east side of Rte. 53 across from the Wal-Mart store south of Butterfield Road (Rte. 56). One lake is used by fishermen; the river looks like the second lake, but it is simply a wide spot in the river with mudflats and cattail marsh. The second lake (a former quarry) is south of the parking lot.

Trails—Walk north from the lot and cross over the stream on the rocks and proceed to the high point overlooking the river and mudflats. (Take care in crossing the stream; the rocks are unstable.) For passerines, walk south over the bridge and follow the path around the quarry lake. The path on the west side of the lake runs between the lake and the river, and the thickets at the south end can be excellent during migration for warblers.

Owned By—DuPage County Forest Preserve District

Telephone—(630)933-7200

Maple Grove Forest Preserve (L16)

These 82 acres of mature oak and maple forest are enriched by a creek and a swampy area. The primary attractions here are woodland migrants. In spring and fall, Carolina Wrens, Winter Wrens, warblers (including Black-throated Blue, Cerulean, Prothonotary, Connecticut, and Hooded Warblers), and many other species have been observed from the bridge and in the swampy section. Among the species known to nest in the preserve are Red-shouldered Hawk, Broad-winged Hawk, Cooper's Hawk, and Tufted Titmouse.

Directions—From I-355, exit east onto Maple Avenue. The main parking lot for Maple Grove Forest Preserve is on Maple Street, but the closest parking lot to the "magic" bridge is on the north side of the preserve. Go east to Lee Avenue (approximately 1.5 miles); turn north to Gilbert Avenue and east to the Gilbert Park parking lot. From I-88, exit south at Highland Avenue. After 3.5 miles, you will cross the Burlington Railroad Tracks. Turn west on the first street after the tracks and continue west on Gilbert for .75 mile to the Gilbert Park lot.

Trails—From the Gilbert Park lot, walk south into the woods toward the bridge over the creek. Birding can be excellent at this bridge. Walk across the bridge and follow the path. The first left leads to a swampy area; other trails traverse the forest and lead to the main parking lot at the south end on Maple Avenue.

Owned By—DuPage County Forest Preserve District

Telephone—(630)933-7200

Lyman Woods (L17)

W S S F

This preserve contains 120 acres of large marsh, pond, fields, and oak woods. Herons and ducks frequent the shallow pond, and numerous migrants utilize the fields and woods.

Directions—Exit I-88 at Highland Avenue southbound and turn east into Braemoor (33rd Street) just south of 31st Street and park along the dead-end road.

Trails—A map of the area is posted in the parking lot. Head south from the parking lot then east to the junction where a trail turns south. The trail can be followed in either direction; it loops back to this point after passing through the marsh and woods. On the eastern portion of the loop, the trail passes through the athletic fields of the Chicago College of Osteopathic Medicine. You must look carefully to find the trail back into the woods. A cement drainage ditch separates the two paths into the woods from the college.

Owned By—Downers Grove Park District

Telephone—(630)963-1300

Fullersburg Woods (L18)

W S S F

Salt Creek runs through these 221 acres of woods. A variety of ducks use this stretch of the creek during spring and fall, but the woods are most fruitful for migrant land birds. White-eyed Vireos, Philadelphia Vireos, and Prothonotary, Connecticut, and Mourning Warblers are just a few examples of what has been seen here. Broad-winged Hawks have nested near the picnic area, and Acadian Flycatchers have summered here.

Directions—Go west of the Tri-State Tollway on Ogden Avenue and turn north at York Avenue. At the stop light, turn left onto Spring Road and look for the entrance to Fullersburg Woods on your right. The preserve can be reached from 31st Street by going south on Spring Road and looking for the entrance on your left.

Trails—Five miles of trails wander through the woods and along Salt Creek. Pick up a trail map at the visitors' center before heading out. Islands in the creek can be reached by crossing the bridge behind the restrooms on the entrance road or by following the interpretive trail north from the visitors' center.

Note—Many programs for schoolchildren are held in the Environmental Education Center here. Naturalists conduct walks and keep a list of bird sightings. To bird when the preserve is quiet, schedule your visit for early or late in the day.

Owned By—DuPage County Forest Preserve District

Telephone—(630)850-8111

Eola Road Marsh (L19)

Also known as **Radio Tower Marsh**, this large wetland is a good spot for waterfowl in the spring and fall. In late May and early June, look for Snowy Egret, Little Blue Heron, Cattle Egret, and rails. In August and September, Forster's Terns and Black Terns may appear. When water levels are low, shorebirds (including Wilson's Phalaropes) are a good possibility. In dry years, the marsh may hold few birds.

Directions—From Eola Road, turn west immediately south of the bridge over the Burlington Railroad tracks. Follow the road north to the dead end and park near the propane filling station or go south and park behind the radio station.

Trails—Walk west to the marsh.

Note—Continue north on Eola Road to another spot for shorebirds and ducks southwest of the intersection of Eola Road and Butterfield Road.

McDowell Grove (L20)

Numerous migrants filter through the woods and brushy areas of this 422-acre preserve, including Olive-sided Flycatchers (which sometimes perch in the treetops south of the bridge) and such unusual warblers as Yellow-throated, Cerulean, Prothonotary, and Connecticut Warblers and Yellow-breasted Chat.

Directions—Exit I-88 at Winfield Road southbound. Turn west on Diehl Road, cross the DuPage River, and turn south onto Raymond Drive. Drive for about one block and turn east into the preserve.

Trails—West of the bridge on the entrance road, take the trail north and east to view islands and a backwater area. East of the bridge, the trail east of the parking area parallels the river and has a good edge for warblers. At the "T" intersection, turn north and follow the river until the trail curves east and back south through meadow and brush. This trail will end up back in the parking lot. Another trail follows the river south of the entrance road to a waterfall which can be excellent for warblers. This trail can be under water or extremely muddy.

Owned By—DuPage County Forest Preserve District

Telephone—(630)933-7200

Springbrook Prairie Forest Preserve (L21)

Springbrook Prairie (along with Fermilab) is one of the best spots for grassland birds in the county. Most of its 1,800 acres are grassy fields, but there are also marshland, a pond, and a short segment of riparian woods. In winter, the grasslands attract Rough-legged Hawks, Long-eared Owls, Short-eared Owls, and Northern Shrikes. The Short-eared Owls are best seen at dusk in the west section of the preserve, west of Book Road and north of 83rd Street. American Woodcock can be observed in spring on the east side of the preserve south of the dog training field. The list of grassland species that have summered is outstanding: Northern Harrier, Upland Sandpiper, Short-eared Owl, Loggerhead Shrike, Bell's Vireo, Sedge Wren, Savannah Sparrow, Grasshopper Sparrow, Henslow's Sparrow, Dickcissel, Bobolink, and Eastern Meadowlark. An excellent variety of birds also use the wetlands at various times of the year: both bitterns, herons, ducks, rails, shorebirds, Sandhill Cranes, Marsh Wrens, and Nelson's Sharp-tailed Sparrows (late September). The woods along Springbrook Creek on the east side of the preserve provide habitat for numerous migrant passerines, including Acadian Flycatcher, White-eyed Vireo, and Connecticut and Mourning Warblers.

Directions—From 75th Street, turn south on Plainfield-Naperville Road and park in the lot on the west side. Parking is also available on the east side of the road in the lot for the dog-training area. The lots do not open until one hour after sunrise.

Trails—The developed trail along Naperville-Plainfield Road is not particularly good for birds. Walk cross-country or follow the footpaths and abandoned roads from the parking lot. Walk west along the marsh and follow the fencerow north to see the wetland near 75th Street.

Owned By—DuPage County Forest Preserve District

Telephone—(630)933-7200

Greene Valley Forest Preserve (L22)

W S S F

Totaling 1,425 acres, this Forest Preserve is richer than most. The East Branch of the DuPage River, coursing through the eastern portion of the preserve, is joined by Anderson Creek from the west. The oak woods north of 79th Street is of exceptional quality, having been zealously protected for over 50 years by the family that owned it. (It is well worth a spring visit for the wildflowers alone.) In addition, the preserve contains grassy fields and evergreens. Long-eared Owls and Northern Saw-whet Owls have wintered in the evergreens, and American Woodcock display in early spring near the Thunderbird Road entrance. During migration, warblers and other passerines prefer the woods along Anderson Creek and the area north of 79th Street. Flycatchers, Yellow-breasted Chats, Savannah Sparrows, and Bobolinks have nested in the fields near the Thunderbird parking lot.

Directions—Drive west of Rte. 53 on 75th Street to Greene Road and turn south. At 79th Street, turn west and park along the road to access the trail along the river or the trail into the woods. For the main parking area, continue .5 mile west to Thunderbird Road and turn south.

Trails—Seven miles of loop trails exist. The trail along the East Branch of the DuPage River east of Greene Road and the trail leading south from 79th Street into the woods just west of Greene Road are good in migration for passerines. Beyond the main parking area, the trail through the youth camp has some pines which should be checked during winter.

Note—Nearby, 83rd Street marsh at Rte. 53 is attractive to shorebirds.

Owned By—DuPage County Forest Preserve District

Telephone—(630)933-7200

Waterfall Glen Forest Preserve (L23)

This preserve owes its size to a program initiated by President Nixon whereby unused federal land would be conveyed to local authorities for use as parks. Under the plan, the federal government gave over 2,000 acres surrounding the Argonne National Laboratory to the Forest Preserve District. The rolling terrain supports a wide variety of habitats, including native woods, evergreen plantings, grassland, marsh, a creek, and even a waterfall. Deadstick Pond is on the west side of the preserve.

The extensive evergreen plantations may have owls (Long-eared and Northern Saw-whet) and crossbills in the winter. In early spring, Eastern Bluebirds return to nest in Poverty Prairie and in re-created oak savannas. Later, breeding Grasshopper Sparrows and Dickcissels join them in the prairie. Other nesters include Red-shouldered Hawks (in the savannas) and Blue-winged Warblers (in the scrubby edges). Herons, Wood Ducks, and Red-headed Woodpeckers favor the environs of Deadstick Pond. And in late September–early October, Turkey Vultures, Ospreys, Northern Harriers, and Sandhill Cranes may be seen migrating overhead.

Directions—Parking is available at three sites. (Do not park inside any gates; they may be locked while you are hiking.) The main lot is on the northeast side of the preserve .25 mile south of I-55 near the Cass Avenue entrance to Argonne National Laboratory. Further south, park along the road where Cass turns east into Bluff Road (99th Street); this is the closest parking area for Sawmill Creek. Access is possible on the west side of the preserve, too, but there are no official parking lots on that side. For Deadstick Pond, turn east from Lemont Road onto Westgate Road just south of I-55 and park on the shoulder of the road near the sign for the Forest Preserve. For Poverty Prairie, turn east from Lemont Road onto South Bluff Road and park on the shoulder of the road.

Trails—A well-marked 8.5-mile trail encircles the Forest Preserve; mowed fire lanes and trail linkages provide additional access. Rocky Glen and the Waterfall are south of Bluff Road (about two miles south of the main parking lot). The trail then follows a bluff along Sawmill Creek for about .5 mile and reaches Poverty Prairie in another 1.5 miles. From Poverty Prairie, it is approximately three miles back around to the main lot; a narrower path parallels the multipurpose trail in two places in the western section of the preserve. From Westgate Road in the northwest part of the preserve, follow the mowed fire lane toward the east instead of the multipurpose trail to view Deadstick Pond. Be wary of speeding bicyclists, especially coming down hills. A compass will be helpful if you take any of the unmarked trails.

Owned By—DuPage County Forest Preserve District

Telephone—(630)933-7200

Kane County

Hampshire Forest Preserve (H1)

W S S F

This 217-acre preserve contains fields, shrubby areas, woods, and conifer plantings. The woods can be good for migrants in the spring and fall, and the extensive grasslands in the vicinity have supported nesting Swainson's Hawks, Savannah Sparrows, Bobolinks, and Western Meadowlarks.

Directions—From I-90, exit at Rte. 47 and go south to Big Timber Road. Turn west on Big Timber and stay on it after it crosses Rte. 20. Turn at the next left, Ketchum Road, which runs along the western boundary of the preserve. Travel south on Ketchum until it ends. Turn east on Allen Road. Just past the small cemetery at that corner is the entrance to the preserve. The entrance drive continues north and east through an open field where the Valley Hawks Soaring Club meets to fly their battery-powered airplanes. Park there for grassland species. The road continues to the north through a wooded area and loops back to the entrance road.

Trails—Equestrian trails traverse the wooded area and continue north to an oak-covered hill.

Owned By—Kane County Forest Preserve District

Telephone—(630)232-5980

Rutland Forest Preserve (H2)

W S S F

Despite being only 58 acres in size, the combination of oaks, marshes, and a small stream attract excellent numbers of migrants.

Directions—Drive west from Elgin on Big Timber Road until you see the sign on the north side of the road. A road circles through the preserve with several parking areas off to the side.

Owned By—Kane County Forest Preserve District

Telephone—(630)232-5980

Map H **Kane County** (facing page)

1. Hampshire Forest Preserve *p. 116*
2. Rutland Forest Preserve *p. 116*
3. Burnidge Forest Preserve *p. 121*
4. Tyler Creek Forest Preserve *p. 121*
5. Trout Park Nature Preserve *p. 118*
6. Otter Creek Bend Wetland Park *p. 121*
7. LeRoy Oakes Forest Preserve *p. 122*
8. Blackhawk Forest Preserve *p. 119*
9. Tekakwitha Woods Forest Preserve *p. 119*
10. Ferson Creek Fen Nature Preserve *p. 119*
11. Norris Woods Nature Preserve *p. 120*
12. Fabyan Forest Preserve *p. 120*
13. Les Arends Forest Preserve *p. 120*
14. Elburn Forest Preserve *p. 122*
15. Johnsons Mound Forest Preserve *p. 122*
16. Nelson Lake Marsh Nature Preserve *p. 123*
17. Bliss Woods Forest Preserve *p. 123*
18. Sugar Grove Marsh *p. 124*
19. Big Rock Forest Preserve *p. 124*

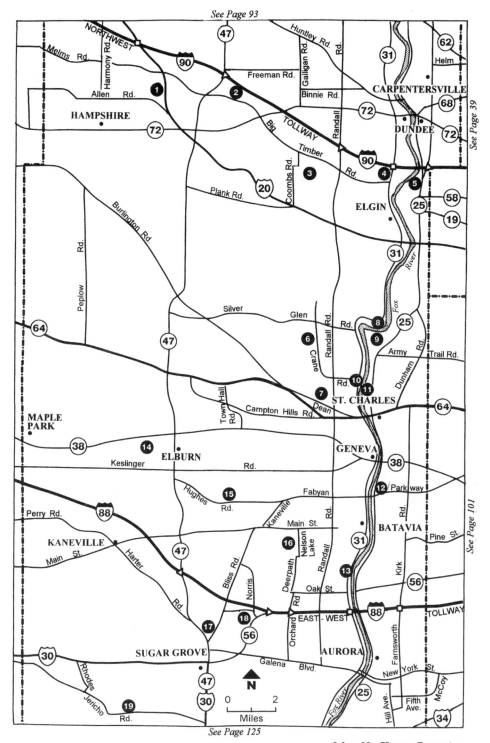

See Page 93

See Page 39

See Page 101

See Page 125

Map H **Kane County**

Fox River Corridor

W S S F

A history of Kane County published in 1904 describes the Fox River with these words: "On the east bank of the river there was a body of magnificent timber, extending from the south line of the county northward to Batavia, called the 'Big Woods,' and a similar growth, reaching from Geneva to the north line of St. Charles township, was called the 'Little Woods.' These tracts were covered with a growth of hardwood trees, standing so thickly and of such stately proportions as to fully justify their designation as timber lands." In Elgin, "there was a small tract of very remarkable forest and plant growth known as the 'Cedar Swamp'" (Bateman et al., 622). And on the river's west side, there were open marshy areas. Remnants of this diversity still survive in preserves owned by the county, municipalities, and even a corporation.

In winter, water birds may be present at areas of open water. Check below the dams at South Elgin, St. Charles, Geneva, Batavia, and North Aurora for such species as Horned Grebe, Wood Duck, Northern Pintail, Bufflehead, Hooded Merganser, Lesser Black-backed Gull, and Glaucous Gull. One year, lucky birders found an immature King Eider. In spring and fall, herons, raptors, flycatchers, vireos, thrushes, warblers, and sparrows can be found in appropriate habitats. Motorboat activity tends to keep the river clear of birds during the warmer months, but Wood Ducks, Broad-winged Hawks, Yellow-throated Vireos, Red-eyed Vireos, Tufted Titmice (permanent residents), Wood Thrushes, Ovenbirds, Scarlet Tanagers, and other woodland birds nest along the corridor.

Directions—Two highways parallel the river and provide access to the various Forest Preserves and dams; Route 31 is on the west side of the river and Rte. 25 is on the east side of the river.

Trails—The 32-mile Fox River Trail passes through many of the parks between Algonquin and Aurora. The trail is on the east side of the river in the northern part of Kane County and crosses to the west side of the river at Fabyan Forest Preserve.

Owned By—The majority of the sites described are owned by the Kane County Forest Preserve District.

Telephone—(630)232-5980

Trout Park Nature Preserve (H5)   Watered by numerous cold seeps rich in calcium, Trout Park (the "Cedar Swamp" referred to earlier) supports a combination of plants and insects that is found nowhere else in Illinois. In recognition of its biological significance, 26 acres of the park have been a dedicated nature preserve since 1972. The perpetually running water and sheltered ravines enable Golden-crowned Kinglets, Yellow-rumped Warblers, and Hermit Thrushes to linger into the winter. But it is the thrushes, vireos, warblers, and other migrants passing through during spring and fall that make this an especially rewarding destination for birders.

Directions—Exit I-90 at Rte. 25. Go south to the park, which is immediately on the right, and proceed to the parking lot near the nature preserve sign.

Trails—A well-marked system of trails, starting at the parking lot, makes the site accessible. Erosion of the steep slopes is a serious problem here, so please stay on the trails currently open. (Some old trails have been blocked for erosion control.)

Owned By—City of Elgin

Blackhawk Forest Preserve (H8)   This wooded preserve of 186 acres is excellent for passerines in migration. Look for sparrows each spring and fall in the shrubby area near the parking area and warblers and vireos in the woods along the river and near the parking lots.

Directions—Exit I-90 at Rte. 31 and go south through Elgin and South Elgin. Rte. 31 curves to the west near the entrance to the preserve.

Trails—Walk north of the lot to the wooded and shrubby areas, along the river and across the river to Tekakwitha Forest Preserve. The riverbank is not very stable here; beware of steep drop-offs. One mile north of the preserve, the dam in South Elgin is worth a look in winter for waterfowl and gulls.

Tekakwitha Woods Forest Preserve (H9)   The property here includes both a Nature Center and wooded ravines along an east/west bend in the river. Herons and raptors (including Osprey) are frequently seen from the bridge during migration, and when the river levels are low, the exposed rocky banks attract sandpipers.

Directions—Go west from Highway 25 on Courier to Illinois, turn south (left) and then west (right) at Villa Marie Road. Entrance to the preserve is on the north side of Villa Marie Road.

Trails—Follow the paved trail north from the parking lot to the Nature Center, which maintains bird feeders. The Fox River is beyond the Nature Center. Trails pass through the woods, parallel the river, and circle back around to the parking lot. East of the parking lot, go north on the Fox River Bicycle Trail to the bridge over the river. This is a good vantage point for waterfowl, raptors, and migrant passerines as you are at treetop level. Blackhawk Forest Preserve is on the north side of the Fox River.

Note—The Nature Center is closed on Fridays and Thursdays from the first of December through the end of March. On weekends, the center is open from noon to 4:00 P.M.; from Monday through Wednesday it is open from 9:00 A.M. to 4:00 P.M.

Telephone—(847)741-8350

Ferson Creek Fen Nature Preserve (H10)   Distinguished by waters laden with calcium and other minerals, fens are among the rarest of wetlands. They are usually fed by springs or by water percolating through morainal uplands. The latter case applies to Ferson Creek Fen, a high quality wetland complex of 46 acres. It is easily traversed via a boardwalk constructed to protect the sensitive vegetation. Winter Wrens and Yellow-rumped Warblers have wintered, and Broad-winged Hawks, Virginia Rails, and Soras have summered. The principal avian attractions, however, are the warblers and other migrants that come through in spring and fall.

Directions—Drive 1.5 miles north of Rte. 64 on Rte. 31. The preserve is north of Ferson Creek Park on the east side of the road.

Trails—A path makes a circle through the preserve. A platform overlooks the Fox River, another overlooks the fen, and a boardwalk borders the floodplain woods.

Owned By—St. Charles Township Park District

Telephone—(630)584-1055

Norris Woods Nature Preserve (H11)   The forest protected within this 73-acre preserve and an adjacent tract of private land is the highest quality example of the original Fox River timber that still survives. Some of the Red Oaks and White Oaks are over 150 years old. Spring wildflowers vie with warblers and other migrants for the birder's attention. Fortunately, the site is particularly good for Connecticut, Mourning, and Canada Warblers, understory species that may pose obligingly in the middle of a bouquet, if only for a moment. The preserve also provides breeding habitat for Wood Ducks, Great Horned Owls, Tufted Titmice, Veeries, Wood Thrushes, and Scarlet Tanagers.

Directions—Turn west from Rte. 25 onto Johner Avenue south of the Anderson Consulting Co. campus and continue west on the road behind Bethlehem Lutheran Church. Park in the lot.

Trails—A trail loops down to the river, up the riverbank, and around through the woods back to the parking lot.

Owned By—St. Charles Township Park District

Telephone—(630)584-1055

Fabyan Forest Preserve (H12)   Two-hundred-forty-five-acre Fabyan Forest Preserve occupies both sides of the river. In fall and winter, finches and siskins may be found in the conifers and gardens near the parking area on the west side. Warblers and vireos are present during migration and seem to prefer the eastern part of the preserve.

Directions—Park in the lot off of Rte. 31 for the west side of the river and in the lot off of Rte. 25 for the east side of the river.

Trails—Walk the trails along the river and check the small area of conifers and oaks near the parking area on the west side of the river.
   Bicycle traffic can be heavy on weekends on the east side of the river.

Les Arends Forest Preserve (H13)   This narrow preserve on the west bank of the Fox River contains 32 acres of woods and thickets. Several wooded islands are visible from the preserve, and an old broken dam upstream creates a series of rapids and riffles. Migrating passerines are common here, and Prothonotary Warblers have nested.

Directions—Exit I-88 at Highway 31 northbound. After passing Moosehart, look for the preserve on the right. Several parking lots and picnic shelters exist.

Trails—The river is very close to the parking lots, and trails extend from the north parking lot toward the river.

Note—**Glenwood Park Forest Preserve**, also productive for migrants, is directly opposite Les Arends on the river's east side. It can be reached off of Rte. 25, two miles south of Batavia.

Burnidge Forest Preserve (H3)

W S S F

Also known as **Paul Wolff Forest Preserve**, this site consists of 486 acres of grasslands, woods, wetlands, and two small fishing ponds. The woods host migrant land birds, and the grasslands support nesting Field Sparrows, Savannah Sparrows, and Bobolinks. Warbling Vireos and Common Yellowthroats nest in the willows around the wetlands.

Directions—The main access to the preserve is from Big Timber Road. During the summer, there is access from Coombs Road on the west, but the gate is closed during most of the year. From the main entrance, drive past the small fishing pond and follow the drive to the west and south, past the primitive camping area to the picnic shelter and one-way drive that circles a hilly area of woods.

Trails—Paths are mowed in the grasslands that separate the campgrounds and picnic area. Trails go south into the large, undeveloped part of the preserve from the wooded area south of the picnic shelter.

Owned By—Kane County Forest Preserve District

Telephone—(847)695-8410

Tyler Creek Forest Preserve (H4)

W S S F

The 57 acres of creek and bottomland woods can be excellent for warblers in spring and fall migration.

Directions—Exit I-90 at Rte. 31. Turn west at the light just south of I-90 to enter the park.

Owned By—Kane County Forest Preserve District

Telephone—(630)232-5980

Otter Creek Bend Wetland Park (H6)

W S S F

Although newly restored, the wetlands here have already attracted Tundra Swans and ducks in early spring and shorebirds in summer and early fall. Flocks of Rusty and Brewer's Blackbirds also use the area during migration.

Directions—Turn west onto Silver Glen Road from Randall Road and then turn south onto Crane Road. Park in the lot on the west side across from the cemetery. The lot has a good overview of the area.

Trails—A circular path surrounds the site and offers an excellent overlook and access.

Owned By—Kane County Forest Preserve District

Telephone—(630)232-5980

LeRoy Oakes Forest Preserve (H7)

W S S F

This 253-acre preserve of mixed habitat is a good spot for migrant flycatchers, warblers, and other passerines. Among the species known to nest here or along the nearby Great Western Trail are White-eyed Vireos, Blue-winged Warblers, Yellow-breasted Chats, and Eastern Towhees.

Directions—Turn onto Dean Street from North Avenue (Rte. 64) or Randall Road. The entrance to the Forest Preserve is on the north side of Dean Street west of Randall Road. Park in the last lot to the north. To reach the Great Western Trail, go south of Dean Street.

Trails—Walk down to the creek at the north end of the preserve.

Owned By—Kane County Forest Preserve District

Telephone—(630)584-5988

Elburn Forest Preserve (H14)

W S S F

This is an 87-acre island of trees and shrubs in a sea of agricultural land. Winter is a good time to look for Rough-legged Hawks, Horned Larks, American Tree Sparrows, Dark-eyed Juncos, Lapland Longspurs, and Snow Buntings in the fields on the east side. During the same season, the woods often harbor Eastern Screech-Owls, woodpeckers, Tufted Titmice, and Brown Creepers. Visits during spring and fall migration should produce vireos, thrushes, warblers, and sparrows.

Directions—Drive west of Rte. 47 on Rte. 38; turn into the preserve on the south side of the road and drive south to the last parking lot.

Trails—Walk east along the railroad tracks and circle around to the north and back to your car.

Owned By—Kane County Forest Preserve District

Telephone—(630)232-5980

Johnsons Mound Forest Preserve (H15)

W S S F

The mound is a timbered kame studded with spring wildflowers. The Shabonna Elm, a tree famed for its enormity, grew here until it was victimized by Dutch elm disease. A good assortment of warblers visit the 185-acre preserve during migration. Broad-winged Hawks, Swainson's Hawks, Barred Owls, and Tufted Titmice (permanent residents) have all summered here or close by.

Directions—Enter the preserve from Hughes Road. There are a few pull-offs along the one-way road that circles through the preserve.

Owned By—Kane County Forest Preserve District

Telephone—(630)232-5980

Nelson Lake Marsh Nature Preserve (H16)

The wetlands of Nelson Lake are unlike any other in the county. A morainal lake, it has gradually filled with peat, leaving a shallow area of open water, fen, and marshes. Wooded uplands add to the diversity of the preserve. Rough-legged Hawk and Northern Shrike are possible in winter in the open areas. Migrant waterfowl are visible from the overlook; walking close to the lake is also worthwhile as it provides a different angle. The fields to the south and west of Nelson Lake are often flooded in the spring and attract shorebirds. In both spring and fall, sparrows gather in the shrubby areas along the trail and near the farm north of the preserve, and vireos and warblers seek insects in the wooded areas along the trail. Connecticut Warblers are seen each year. Virginia Rails, Soras, Common Moorhens, Sandhill Cranes, American Redstarts, and Yellow-headed Blackbirds have all summered in the marsh, and terns kite over the water.

Directions—Park in lot on the west side of Nelson Lake Road about .5 mile south of Main Street. This lot is elevated above the level of the lake and affords a nice overview.

Trails—Take the trail north out of the lot. The first ponds on the left are shallow and often have Soras. After passing a deeper pond, the trail veers west and south along ponds and marshes. The trail then goes up a hill, through woods, and to a platform from which you can view the lake.

Owned By—Kane County Forest Preserve District

Telephone—(630)232-5980

Bliss Woods Forest Preserve (H17)

Raspberry Creek curves through the 280 acres of deep woods, wetlands, and fields that comprise this preserve. The site is probably best known for its spring displays of Dutchman's Breeches, trilliums, and other flowers, but birding can also be very rewarding during migration. Vireos, thrushes, warblers, and other passerines move through in good numbers. Carolina Wrens are noteworthy summer residents.

Directions—Exit I-88 at Rte. 56 toward Sugar Grove. Turn west on Galena Boulevard to Rte. 47. Turn north on Rte. 47, then right on Bliss Road. Look for the entrance to the preserve after .4 mile on the west side of the road.

Trails—Head north on the trail that circles the preserve. There are many trails through this former Boy Scout Camp, including the paved Virgil Gilman Trail which cuts through the preserve from east to west. Bicycle traffic may be heavy on this trail. To see the eastern portion of the preserve, cross Bliss Road and walk east on the paved trail. Return to the parking area by going through the camping area. The sounds of trap shooting from the Aurora Sportsmen's Club across the street can be disquieting.

Owned By—Kane County Forest Preserve District

Telephone—(630)232-5980

Sugar Grove Marsh (H18)

W S S F

Also known as Denny Road Marsh, this shallow lake and marsh is in the middle of agricultural fields. Greater White-fronted Geese and Bonaparte's Gulls have occurred in spring, while Common Moorhens, Sandhill Cranes, and Yellow-headed Blackbirds have nested. Cormorants, herons, and ducks also use the area.

Directions—From Bliss Road, turn east on Denny Road and proceed until you see the marsh on the south side of the road. The road curves to the south where there is a better view. This marsh is on private property and must be viewed from Denny Road.

Trails—Bird from the road. Binoculars are sufficient, but a spotting scope is helpful.

Big Rock Forest Preserve (H19)

W S S F

The heart of this 298-acre preserve is a rich bottomland woods along the east bank of Big Rock Creek. Sycamores grow here, as do Illinois's northernmost native stands of Redbud and Kentucky Coffee trees. Migrant passerines are common, and nesting species include Barred Owls, Tufted Titmice, Wood Thrushes, Ovenbirds, and Scarlet Tanagers.

Directions—Exit I-88 at Rte. 56 and turn south on Rte. 47. After two miles, turn right onto Jericho Road and proceed three miles. As you approach the creek, look for the small white sign with black lettering identifying the property as a Kane County Forest Preserve. There is a pull-off on the north side of the road east of the creek, or you can cross the creek and turn south on the gravel road which parallels the creek on the west side.

Trails—A trail heads north from the pull-off on the north side of Jericho Road. South of the road, you can see the creek area from the gravel road.

Owned By—Kane County Forest Preserve

Telephone—(630)232-5989

Kendall County

Silver Springs State Park (N1)

W S S F

The diversity of habitat within the state park makes it the best birding area in the county. There are woods, grasslands, lakes, marshes, and a stretch of the Fox River. During the winter, fields south of the park office often have Northern Harriers, Horned Larks, Lapland Longspurs, and Snow Buntings. In spring and fall, Buffleheads, Common Goldeneyes, all three mergansers, and other waterfowl may be seen on the river. Ospreys and numerous passerines also frequent the park during migration. Nesting species in the main part of the park include American Woodcock, Willow Flycatcher, Warbling Vireo, Wood Thrush, Yellow-breasted Chat, and Orchard Oriole. The Duck Creek Nature Trail area has nesting

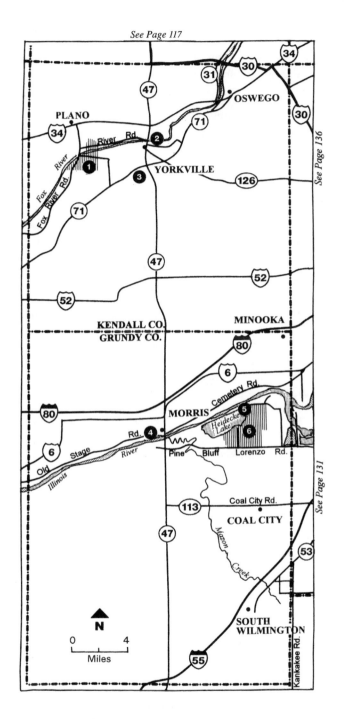

Map N   **Kendall and Grundy Counties**

Kendall County

1. Silver Springs State Park *p. 124*

2. Yorkville Dam *p. 126*

3. Harris Forest Preserve *p. 127*

Grundy County

4. Gebhard Woods State Park *p. 127*

5. Heidecke Lake State Fish and Wildlife Area *p. 127*

6. Goose Lake Prairie State Natural Area *p. 129*

Red-shouldered Hawks, Barred Owls, Acadian Flycatchers, Yellow-throated Vireos, Tufted Titmice, Ovenbirds, and Scarlet Tanagers. The Prairie Restoration area has breeding Grasshopper Sparrows, Bobolinks, and Orchard Orioles. The Kendall County Outdoor Education Center is located in the state park.

Directions—From Plano and Rte. 34, turn south on Ben Street (Fox River Road) and follow the signs to the park. From Yorkville, drive west on River Road on the north side of the river. The main entrance is south of the river on Fox Road and provides access to two fishing lakes, a pond, the Fox River, and Silver Springs. For Duck Creek Nature Trail, go west of the main entrance toward Cedar Ridge Primitive Campground, go to the end of the road, and park in lot #4. For the prairie restoration area, go north on Fox River Road across the river, turn east on River Road, and park in lot #5 after .5 mile. The park office and the Kendall County Outdoor Education Center are on Fox Road.

Trails—Trails are located in several sections of the park. From the lot near the park entrance, walk east on the trail along the river for about .1 mile. Go inland to walk along the small lake and continue east and south to the swampy area near Silver Springs. Retrace your steps to your car or continue on this trail (The Fox Ridge Trail) for 3.5 miles as it loops through various habitats and ends up back at the parking lot. The 1.2 mile Duck Creek Nature Trail crosses a small creek and deep bottomland woods. In warm weather, fishermen and picnickers crowd the park after mid-morning, but Duck Creek Trail, the Prairie Restoration area, and the eastern part of the Fox River Trail are usually not crowded. Much of the park is closed during hunting season.

Owned By—Illinois Department of Natural Resources

Telephone—(630)553-6297

Yorkville Dam (N2)

W S S F

Because this section of river tends to stay unfrozen, it is one of the best sites in the county for winter birding. Pied-billed Grebe, American Black Duck, Gadwall, Northern Pintail, Green-winged Teal, American Wigeon, Common Merganser, Common Goldeneye, and rare gulls have all been seen here. In addition to the open water, check the gravel bars and island below the dam. In the spring and summer, look for Black Terns and Rough-winged Swallows.

Directions—Drive east of Rte. 47 on Hydraulic Avenue in Yorkville just south of the river and park in two blocks at the city park.

Trails—View the dam from the park.

Note—The river can also be viewed from **Saw Wee Kee Park**, 5 miles east of Yorkville. A Bald Eagle was seen here one winter. The park's woods and ponds provide habitat for resident Carolina Wrens and migrant herons, waterfowl, vireos, thrushes, and warblers. The park is 1 mile north of Rte. 71 on Minkler Road in Oswego. Drive to the far west end and park in the gravel lot. The ponds can be found by walking west on the dirt road for approximately 100 yards and taking the trail to the south. The park is owned by the Oswegoland Park District; telephone—(630)554–1010.

Harris Forest Preserve (N3)

This 85-acre park has woods, a woodland pond, recreational fields, a toboggan hill, and a fishing lake. It attracts many woodland migrants, as well as nesting Eastern Wood-Pewees and Wood Thrushes.

Directions—Drive .5 mile west of Rte. 47 on Rte. 71. Turn south into the park and follow the loop through the woods. There are several areas to park along the road.

Trails—Walk the nature trail.

Owned By—Kendall County Forest Preserve District

Telephone—(630)553-4025

Grundy County

Gebhard Woods State Park (N4)

Within Grundy County, the Illinois and Michigan (I & M) Canal Trail follows the north bank of the canal from the eastern boundary of the county to the western. Portions of the route are adjacent to the Illinois River, which often has ducks, gulls, and terns in season. One of the trail's best stretches for birders lies along the southern edge of Gebhard Woods State Park, a small stand of sycamores, cottonwoods, and maples. The park can be excellent for migrating wetland birds, thrushes, and warblers. Birding is also good for woodland migrants in the area extending from the park to the county's western boundary.

Directions—Exit south off I-80 onto Rte. 47. Travel south to Jefferson Street in the town of Morris, and turn west. This street becomes Fremont Street. At Ottowa, turn south (left), then turn left as soon as you cross the bridge over Nettle Creek and park in the lot for the visitors' center.

Trails—To bird Gebhard Woods State Park, walk the Nettle Creek Nature Trail and the towpath along the I & M Canal.

Owned By—Illinois Department of Natural Resources

Telephone—(815)942-0796

Note—Other parking areas allowing access to the I & M Canal Trail are located at Aux Sable Preserve, William G. Stratton State Park, along Old Stage Road, or in the town of Seneca (in LaSalle County, just west of the Grundy County line).

Heidecke Lake State Fish and Wildlife Area (N5)

Located between the Illinois River and Goose Lake Prairie State Park, Heidecke Lake is an 1,800-acre cooling lake that the Department of Natural Resources (DNR) leases from Commonwealth Edison for fishing and hunting. It is the single most reliable location in the region for Bald Eagles, two or three of which usually spend the winter. Other wintering birds that have taken advantage of

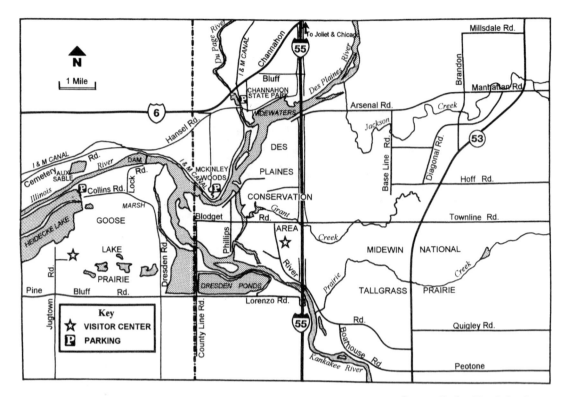

Goose Lake Prairie Area

the open water include White-winged Scoter, Oldsquaw, Common Goldeneye, Thayer's Gull, Iceland Gull, Lesser Black-backed Gull, Glaucous Gull, and Great Black-backed Gull. Ring-billed Gulls nest at the west (private) end. In fall, Black-bellied Plovers, American Golden-Plovers, and Buff-breasted Sandpipers have been seen on the rocky dikes.

Directions—Exit I-55 at Lorenzo Road. Proceed west to Dresden Road and turn north. Follow the road as it curves to the west and ends at Heidecke Lake and park in the lot. Another point of access, gated during the winter, is further west at the northern end of Jugtown Road. Access to the far west end of the lake is restricted by the Commonwealth Edison power station, but many rarities have been seen at that end.

Trails—It is possible to bird from your car at the east end. During fishing season, you can walk the dikes.

Owned By—Commonwealth Edison

Telephone—(815)942-6352

Note—Nearby **Dresden Lock and Dam**, situated on the Illinois River, is also an excellent place to see waterfowl in the winter. From I-55 go west on Lorenzo Road and turn north on Dresden Road. (Shrubby areas near the intersection of the two roads have had summering White-eyed Vireos, Bell's Vireos, and Yellow-breasted Chats.) Continue on Collins Road as it curves to the left. Take Collins to Lock Road and turn north to the parking lot for the lock and dam. There is a heated bathroom here. The lock and dam are also visible from the I & M Canal Trail on the north side of the river.

Goose Lake Prairie State Park (N6)

W S S F

A traveler in 1861 described the prairie near Chicago this way: "The prairies are to my notion impressive at all times. . . . There is something inexpressibly grand in their huge, boundless extent, their gently waving mounds and thick, long grass, which make the distant horizon wavy and indistinct as that of the ocean itself" (Wood).

Back then, the prairie that was to become this park included a 1,000-acre lake that would be so covered with geese and ducks that the water was no longer visible. That lake and the birds gave this site its name. Although Goose Lake has since been drained and its clay turned into pottery, the grassy expanse that remains is the only prairie in the region big enough to still convey the feeling of grandeur that struck that nineteenth-century writer so forcefully. Although some parts of the area had been mined for coal and other parts grazed, the diversity and size of Goose Lake Prairie made saving it worth fighting for. The ensuing battle for preservation, waged in the late 1960s, was to change the way people think about these marvelous grasslands.

The state park has 2,350 acres, of which 1,537 acres are dedicated as an Illinois Nature Preserve. Perhaps because it is so big and diverse, Goose Lake Prairie is home to two species of root-borer moths *(Papapaipema)* that are among the rarest animals in the region. One has a global range consisting of two prairies, and the other is known from only four prairies. Much easier to find, though, are the numerous birds that inhabit the site. In winter, Northern Harriers, Rough-legged Hawks, Short-eared Owls, and other raptors work the grasslands rich with rodents. With the coming of spring, Common Snipe and American Woodcock begin their courtship flights while American Golden-Plovers forage in flooded fields south of the park. Waterfowl also stop where there is sufficient water. Most birders, however, visit Goose Lake Prairie in late spring and early summer to see the many nesting species. These have included American Bittern (in the marsh at the northern end of the park), Least Bittern, all three rails (near the boardwalk), Northern Bobwhite, Loggerhead Shrike (near the picnic areas), Bell's Vireo, and Yellow-breasted Chat. Goose Lake Prairie has long been recognized as being the most reliable place in northeastern Illinois to find summering Henslow's Sparrows (near the Tallgrass Nature Trail).

Directions—Exit I-55 at Lorenzo/Pine Bluff Road. Drive west seven miles and turn right on Jugtown Road to go to the park office and interpretive center. To reach the northern sections, drive north on Dresden Road, then west on Collins Road, and view the marsh on the south side of the road before the intersection of Lock Road to Dresden Lock and Dam.

Trails—7.5 miles of trails have been established in the park. The Tallgrass Nature Trail and Marsh Loop trails east of the visitors' center total 4 miles. The nature trail passes through prairie and potholes and circles back toward the visitors' center past a pond and log cabin, or connects to the Marsh Loop with its floating bridge. Beyond the Marsh Loop, the Sagashka Trail with its photo blind has good habitat for herons and egrets.

Owned By—Illinois Department of Natural Resources

Telephone—(815)942-2899

Note—The interpretive center is closed on weekends from December through March, Christmas, and New Year's Day. The rest of the year, it is open from 10:00 A.M. to 5:00 P.M. on weekends and 10:00 to 4:00 on weekdays. The road on the east side of the park, Dresden-Collins Road, can have interesting birds. Good places to look include the alders near Lorenzo Road (finches in winter), the marshy ditch farther north on the road, shrubby areas near the intersection of Collins Road and Dresden Lock Road (White-eyed Vireos, Bell's Vireos, and Yellow-breasted Chats in summer), and the power plant canal (Cliff Swallows).

Will County

Channahon State Park (R1)

W S S F

Completed in 1848, the I & M Canal provided an unobstructed waterway connecting the Mississippi River with the Great Lakes via the Illinois River. It was an achievement of singular importance to the early growth of Chicago. Congress recognized the canal's significance when they designated it as the first National Heritage Corridor in 1984. This small park on the canal contains large trees that host numerous spring landbirds. Water birds use the canal in migration and during winters when the water remains unfrozen. Belted Kingfishers are usually present.

Directions—Exit I-55 at Rte. 6 westbound. Look for signs for the park. Turn south on Center Street and west on Storey Street. The park is at the end of Storey Street.

Trails—Walk along the tow path. An unmarked trail heads east from the campground for 3.8 miles to I-55, and the I & M Canal heads west for 52 miles. McKinley Woods is 3 miles west; Dresden Lock and Dam is 5.8 miles west.

Owned By—State of Illinois

Map R **Western Will and Kankakee Counties** (facing page)

Will County

1. Channahon State Park *p. 130*
2. Widewaters *p. 132*
3. McKinley Woods Forest Preserve *p. 132*
4. Dresden Ponds *p. 132*
5. Des Plaines Conservation Area *p. 133*
6. Midewin National Tallgrass Prairie *p. 133*
7. Braidwood Dunes and Savanna Nature Preserve *p. 134*

8. Sand Ridge Savanna Nature Preserve *p. 135*
9. Braidwood Fish & Wildlife Area *p. 135*

Kankakee County

10. Kankakee River State Park *p. 144*
11. Gar Creek Trail and Prairie *p. 144*
12. Iroquois Woods Natural Area *p. 144*

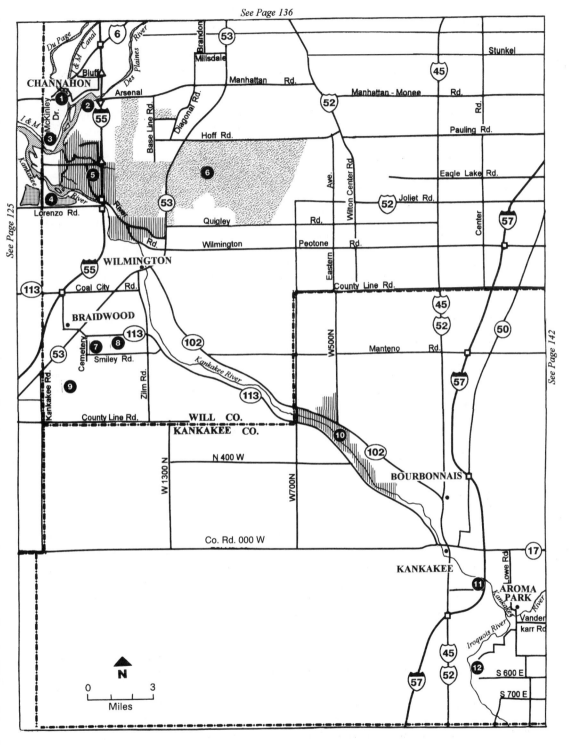

See Page 136

See Page 125

See Page 142

Map R **Western Will and Kankakee Counties**

Widewaters (R2)

W S S F

This swelling of the Des Plaines River east of the town of Channahon can be very good for diving ducks and gulls in winter. Herons and dabbling ducks are seen from spring to fall, and occasional shorebirds stop during migration.

Directions—Exit I-55 at Bluff Road westbound, go south on Frontage Road to the marina area, turn west on Front/River Road, and follow the road along the river.

McKinley Woods Forest Preserve (R3)

W S S F

Migrant warblers and other land birds are partial to this 473-acre wooded preserve along the Des Plaines River. One particularly good spot is the lower parking lot. Summering species include Turkey Vulture, Broad-winged Hawk, Brown Creeper, Prothonotary Warbler, Ovenbird, and Kentucky Warbler.

Directions—Exit I-55 at Rte. 6, go west 3.5 miles, then turn sharply left onto Bridge Street and immediately south on McKinley Drive to the park. Park in the lower parking lot.

Trails—Trails circle through the preserve. The 60-mile long I & M Canal State Trail is accessible here. Walk three miles east to Channahon State Park on the trail or west all the way to the town of LaSalle.

Note—The preserve gate is locked at 5:00 P.M. from November through March and at 8:00 P.M. other times of year. The steep road to the lower parking lot may be closed if conditions are icy.

Owned By—Will County Forest Preserve District

Telephone—(815)727-8700

Dresden Ponds (R4)

W S S F

A 10-mile system of dikes shunts water between the Dresden Nuclear Station and cooling ponds. From November to April, the unfrozen ponds are magnets for such ducks as American Wigeon, American Black Ducks, teal, Northern Pintail, Canvasbacks, Redheads, Ring-necked Ducks, Greater Scaup, mergansers, and Ruddy Ducks. An American White Pelican once spent the winter. A colony of Ring-billed Gulls nests on the east-west dike in the east pond and can be seen from Lorenzo and Cottage Roads.

Directions—Exit I-55 at Lorenzo/Pine Bluff Road and go west. In 1.5 miles, the largest cooling pond will be visible on your right. Turn north on Cottage Road to view the east end of the lake or proceed west another 1.5 miles to County Line Road and turn north to view the west end of the lake. Ponds will be on both sides of County Line Road, and there is room to pull over and park before passing through a covered bridge. Fog can form on cold days and make visibility quite poor. Bird from the bridges over the ponds or from Cottage Road.

Owned By—Commonwealth Edison

Des Plaines Conservation Area (R5)

Between the Des Plaines and Kankakee Rivers, the Illinois Department of Natural Resources maintains these 5,080 acres of fields, thickets, woods, and marshes. Winter land birding consists of scanning open areas for such species as Rough-legged Hawk, Northern Bobwhite, and Eastern Meadowlark. Water birds present in winter have included Black-crowned Night-Heron, Gadwall, American Wigeon, Ring-necked Duck, Lesser Scaup, Bufflehead, Common Goldeneye, Hooded Merganser, and Killdeer. Three Rivers Marina is a good spot to check the river. During the spring and fall migration periods, look for waterfowl, shorebirds, and passerines. But the most interesting birding is in spring and summer for nesting species. Breeding birds have included Upland Sandpiper, both cuckoos, Whip-poor-will, White-eyed Vireo, Bell's Vireo, Tufted Titmouse, American Redstart, Prothonotary Warbler, Yellow-breasted Chat, Summer Tanager, Vesper Sparrow, Henslow's Sparrow, Western Meadowlark, and Orchard Oriole. Some of the better areas to cover include: (1) the grassland and thickets west of Frontage Road, (2) the pond at Blodgett Road and Frontage Road, (3) where Grant Creek crosses Blodgett just west of I-55, (4) the fields south of headquarters, (5) the area around the stables and Field Trial headquarters, and (6) **Grant Creek Prairie**.

Directions—Get off I-55 at exit 241 (River Road) on the north side of the Kankakee River. Most of the Conservation area is west of I-55, but 78-acre Grant Creek Prairie is along the east side of I-55. Drive west and north on River Road to the park headquarters. North of the park office, take Blodgett Road to the west for a view of the Grant Creek cutoff. Continue west to Phillips Road and north to the Three Rivers Marina. East of River Road, take Blodgett to the frontage road on the west side of I-55. Go north to the gravel road and follow it west to the wildlife refuge area of grassland and thickets. The southern portion of the Conservation Area is reached by going south on River Road past I-55 and south on Boathouse Road to the Field Trial Headquarters and Stable.

Trails—On the west side of I-55, bird the fields south of the park headquarters, the Grant Creek cutoff, Three Rivers Marina, the pond at Blodgett and Frontage Roads, and the grassland and thickets north of Blodgett and west of Frontage Road. East of I-55, check the Grant Creek Prairie natural area north of the River Road turnoff, the Kankakee River, and the fields across Boathouse Road from the Field Trial Headquarters and Stable. The park office is closed in winter; hunting is allowed, and roads may be closed for this activity.

Owned By—Illinois Department of Natural Resources

Telephone—(815)423-5326

Midewin National Tallgrass Prairie (R6)

When the federal government decided to decommission the Joliet Army Arsenal, conservation forces led by the Illinois Department of Natural Resources fought to ensure the preservation of the area. Their success in creating Midewin, the first national grassland east of the Mississippi River, means that 19,000 acres, overwhelmingly the largest tract of publicly owned land in the region, will be protected from development. Public access is currently limited, so bird from the road or call the park office in advance to arrange a tour. Eventually, a system of trails will be established to allow greater access. This immense area supports the largest population of nesting Upland Sandpipers in the entire state of Illinois and the largest population of nesting Loggerhead Shrikes in northern Illinois. Numerous other species also utilize this unique site. In winter, birders may find Northern Harriers, Rough-legged Hawks, Long-eared Owls, and Short-eared Owls. Migration brings sparrows to the open areas and warblers to the woodlands. Birding is probably best, however, in spring and summer, for there is a wonderful array of species known to breed here. In addition to the birds previously mentioned, Pied-billed Grebes, Northern Harriers, Hooded Mergansers, Common Snipe, Common Moorhens, King Rails, Acadian Flycatchers, Least Flycatchers, Northern Mockingbirds, Bell's Vireos, Cerulean Warblers, American Redstarts, Kentucky Warblers, Mourning Warblers, Grasshopper Sparrows, Blue Grosbeaks, Dickcissels, Bobolinks, and Western Meadowlarks have all nested here.

Directions—Exit I-55 eastbound at Exit 245 (Arsenal Road). The Joliet Army Training Area is north of this area and can be birded from Brandon and Millsdale Roads. For the areas further south, go east on Arsenal Road for one mile to Base Line Road, turn south for one mile and then turn east. This stretch through grasslands is good for Upland Sandpipers and Loggerhead Shrikes. Turn south on Diagonal Road, then east on Hoff Road to Highway 53. Proceed two miles south, then turn west and bird Employee Gate 10 road and Central Road west of Highway 53. Further south on Highway 53, the U.S. Forest Service Headquarters is located in a house at 30071 South Rte. 53, one mile north of River Road. The southern boundary of the arsenal is on South Arsenal-Quigley Road. To return to I-55 northbound, go west on River Road.

Trails—Trails will be developed in the future. At present, it is necessary to register for weekend tours or to bird from the road.

Owned By—U.S. Forest Service

Telephone—(815)423-6370

Braidwood Dunes & Savanna Nature Preserve (R7)

W S S F

The basin once covered by glacial Lake Morris was an extensive plain of sand ridges and wet swales. Throughout the growing season, the panorama would change as first one and then another flower would burst into color. Poorly suited for agriculture, portions of the area remained virtually uninhabited by people until well into the 1920s and beyond. In the subsequent decades however, development did erase most of the original landscape, and this 288-acre parcel, owned for many years by the Peabody Coal Company, is among the best surviving examples. Most birders cover the area in spring and summer as they search for breeding species of southern affinity: Northern Bobwhite, Whip-poor-will, Bell's Vireo, Northern Mockingbird, Summer Tanager, Lark Sparrow, Henslow's Sparrow, Blue Grosbeak, Dickcissel, and Orchard Oriole.

Directions—Exit I-55 at Rte. 113 (Exit 236). Stay on Rte. 113 as it goes east over the bridge, turns south for 1.9 miles, turns east at a stop sign, and curves right as it crosses Rte. 129 and Rte. 53. Park in the lot .9 mile east of Rte. 53 on Rte. 113.

Trails—Walk the 1.5 miles of trails south of the parking lot off of Rte. 113. To walk in from the south, turn south at Reed (Cemetery) Road at the high school. After one mile, drive east on unmarked Smiley Road for .25 mile, park on the north side, and walk north.

Note—The avian attractions of the 200-acre **Sand Ridge Savanna Nature Preserve** (R8) are similar to those of Braidwood Dunes and Savanna. Sand Ridge is 1.5 miles east of Braidwood Dunes and 1 mile east of Essex Road, on the south side of the road. Birding can be good from the roadside.

Owned By—Will County Forest Preserve District

Telephone—(815)727-8700

Braidwood Fish & Wildlife Area (R9)

The Illinois Department of Natural Resources leases from Commonwealth Edison the large lake used to cool water for the company's nuclear power plant. This lake, which is heavily used by anglers, comprises the bulk of the wildlife area. Wintering wildfowl take advantage of the warm water, and Ospreys are seen regularly in migration. Northern Bobwhite, Northern Mockingbirds, Lark Sparrows, and Orchard Orioles nest in the adjacent fields.

Directions—Birders generally approach the lake from the north. Turn south from Rte. 113 onto Cemetery Road (Reed Road) and drive to the end of the road. There is a parking lot for fishermen on the west and a cemetery on the east. The lake can also be seen from Kankakee Road on the west and south shores of the lake.

Trails—For land birds, look along the edge of the cemetery.

Will County Sod Farms (O1)

Although residential development whittles away at these fields, the sod farms that persist retain their attractiveness to southbound shorebirds. Look for Upland Sandpipers in early August, Buff-breasted Sandpipers from mid-August to early September, and American Golden-Plovers in September. They tend to concentrate where the sod has either been freshly watered or recently stripped. During the same time period, Cattle Egrets have frequented the pastures at 95th and Hegg Roads. One interesting species that breeds in the area is Cliff Swallow. In recent years, a colony of them has nested under the bridge on 127th Street at the DuPage River east of Rte. 59.

Directions—From the northwest corner of Will County, drive south and east along the back roads searching for birds from your car. Look for flooded fields or sod farms where the sod has been stripped away between 91st Street on the north and 127th Street on the south on either side of Rte. 59. To observe with your scope, park at the school located on the northwest corner of Rte. 59 and 103rd Street or at the **Naperville Polo Club** further south.

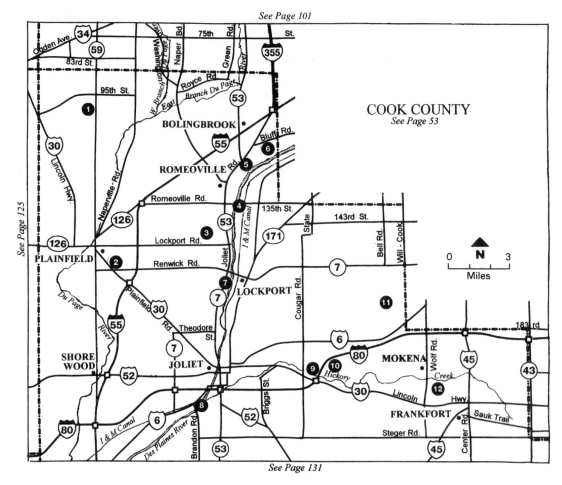

See Page 101

COOK COUNTY
See Page 53

See Page 131

Map O **Northern Will County**

1. Will County Sod Farms *p. 135*
2. Lake Renwick Heron Rookery Nature Preserve *p. 137*
3. O'Hara Woods Nature Preserve *p. 137*
4. Isle a la Cache *p. 137*
5. Veteran's Woods Forest Preserve *p. 138*
6. Keepataw Preserve *p. 138*
7. Lockport Prairie Nature Preserve *p. 139*
8. Brandon Lock and Dam *p. 139*
9. Pilcher Park *p. 140*
10. Higginbotham Woods *p. 140*
11. Messenger Woods Nature Preserve *p. 140*
12. Hickory Creek Preserve *p. 141*

Trails—Most birding here is done from the roadsides. If you park at the school, walk to the north edge of the school property to view the fields to the north.

Note—The Naperville Polo Club grounds often host the same group of shorebirds that are found at the sod farms across the street. Birds will move from one site to the other depending on such circumstances as moisture conditions and whether a match is in progress (which is often the case on Saturdays and Sundays). Turn east from Rte. 59 onto 119th Street. Park at the polo clubhouse lot on the north side of the street and bird from the parking lot.

Lake Renwick Heron Rookery Nature Preserve (O2)

Quarrying operations created this 320-acre pit now filled with water. Great Egrets and Black-crowned Night-Herons have nested here since at least 1961. Two years later Great Blue Herons arrived. In the intervening years, these species have been joined by Double-crested Cormorants and Cattle Egrets. With hundreds of birds nesting on its four principal islands, Lake Renwick has become the premier heron rookery in the region. The herons return each spring in March/April and are present through September. American White Pelicans, ducks, terns, and other migrant water birds also use the lake.

Directions—Exit I-55 at Rte. 30 and go northwest toward Aurora. Turn right at Renwick Road to the visitors' center or park along Rte. 30 north of Renwick Road and bird from the roadside. The visitors' center is open on Saturday mornings in May and June from 8:00 A.M. to 12:00 noon. It is also open on Wednesdays at 10:00 A.M. for a one-hour tour.

Owned By—Will County Forest Preserve District

Telephone—(815)727-8700

O'Hara Woods Nature Preserve (O3)

The Illinois Nature Preserves Commission (1991, 241) says that O'Hara Woods "is one of the last undisturbed maple prairie groves remaining in northeastern Illinois." Marshy ground and a deep ravine held the fires at bay, thus enabling the rich forest to develop. This is yet another place where the birder in spring will be torn between looking at warblers in the canopy above and the pucoon, Virginia Bluebells, Squirrel Corn, and Blue-eyed Mary adorning the forest floor. The field near the parking lot occasionally floods, at which time it can be very good for shorebirds.

Directions—Go west of Highway 53 on Romeoville Road for one mile, then turn and go north .3 mile. The nature preserve is to the west.

Trails—A network of trails has been established here.

Owned By—Village of Romeoville

Telephone—(815)886-6222

Isle a la Cache (O4)

Spring and fall birding is often good on this 87-acre island of woods and marsh in the Des Plaines River. Summering species have included Common Moorhens, Brown Creepers, and Prothonotary Warblers. The museum focuses on the region's early French period and is well worth visiting.

Directions—Go east of Rte. 53 on Romeo Road (135th Street) to the lot for the museum. The preserve is open from 8:00 A.M. to 8:00 P.M. (5:00 November–March), and the museum is open from 10:00 to 4:00.

Trails—6 miles of hiking trails (part of the Centennial Trial) extend to the north on the east side of the river.

Owned By—Will County Forest Preserve District

Telephone—(815)727-8700

Note—**Romeoville Prairie Nature Preserve**, directly across Romeo Road from Isle a la Cache, is a 251-acre prairie and wetland rich in unusual plants including the federally endangered Leafy Prairie Clover. For birders, it offers nesting Virginia Rails, Soras, Sedge Wrens, and Swamp Sparrows.

Veteran's Woods (O5)

W S S F

The combination of woods, Des Plaines River, and ravines makes this 77-acre preserve an excellent location for woodland migrants, including numerous warblers.

Directions—There are two parking lots for the preserve on Joliet Road south of Bluff Road. The first lot on the north is for Traders' Corner with a large mowed grass field. The next parking lot is for Acorn Grove which is closer to the ravines and wooded areas. The lots are open from 8:00 A.M. to 4:00 P.M.

Trails—A .25-mile trail through the woods connects the two parking areas, and smaller, unofficial trails lead south down the hill. From the parking lot for Acorn Grove, a small creek can be found by walking beyond the picnic shelter.

Owned By—Will County Forest Preserve District

Telephone—(815)727-8700

Keepataw Preserve (O6)

W S S F

This 216-acre preserve has an extraordinary mix of habitats: forest, bluffs, ravines, Des Plaines River, floodplain, marshes, springs, and grassy ponds. It is a haven for rare plants and animals, including some that are federally threatened or endangered. During spring and fall, landbird migrants can be abundant. Herons wade along the grassy ponds in summer, and Red-shouldered Hawks are likely nesters.

Directions—Go one mile east from Joliet Road on Bluff Road to the sign for the preserve on the south side of the road. The lot is closed on weekends to prevent the preserve from being used for drinking parties that plague nearby **Black Partridge Preserve**.

Trails—A .25-mile trail leads to an overlook of the Des Plaines River Valley.

Note—The parking lot is open during the week only from 8:00 A.M. to 5:00 P.M. and closed on weekends. Rangers patrol the preserve on a regular basis.

Owned By—Will County Forest Preserve District

Telephone—(815)727-8700

Lockport Prairie Nature Preserve (O7)

Within its 254 acres of very high quality prairie and wetland, scattered trees, ponds, and Des Plaines riverfront, Lockport Prairie harbors a host of rare plants and animals. The Lakeside Daisy, once native to the area until the last population was inadvertently buried by a truckload of coal, has been reestablished, and the federally endangered Leafy Prairie Clover can be seen blooming in late summer. As for birds, occasional Great Blue Herons and ducks are found along the river in winter. The ponds have Pied-billed Grebes, Great Egrets, Blue-winged Teal, Northern Shovelers, and Ring-necked Ducks in migration. In summer, look for herons (including Yellow-crowned Night-Herons) when the water is low. Least Bitterns and King Rails have also occurred. White-eyed Vireos, Bell's Vireos, Yellow-breasted Chats, and Orchard Orioles have nested in thickets here, but restoration activities have removed much of the habitat utilized by these species. Carolina Wrens inhabit the woods along the river.

Directions—Turn east from Rte. 53 onto Division south of Highway 7. This road is gated and ends at the river. Park along the road. For a view of the river, go to the cement bridge north of Highway 7. The Lockport Locks are directly east of the prairie and must be approached from the town of Lockport. To view the lock, turn south of Highway 7 on the east side of the river in Lockport and turn west on Division Street. Drive west and then south along the canal.

Trails—In the prairie nature preserve, you can see many birds from your car, and a trail heads south from Division through the prairie to a grove of trees. For the river, scope north and south of the cement bridge. On the east side of the river the drive parallels the canal on a dike which affords a bird's-eye view of the trees to the west and canal below.

Note—The entrance road to the prairie is open from 8:00 A.M. to 5:00 P.M. (8:00 P.M. in summer).

Owned By—Metropolitan Water Reclamation District of Greater Chicago, Will County; managed by Will County Forest Preserve District.

Telephone—(815)726-3306

Brandon Lock and Dam (O8)

Open water persists below this lock and dam on the Des Plaines River during the winter and attracts occasional herons and flocks of ducks. During migration, ducks, gulls, terns, and, if the water is low enough, shorebirds visit the site.

Directions—Exit I-80 at Highway 7 southbound. Go east on Rte. 6 and check the river above the dam, then turn back and go south on Brandon Road. Park along the bridge. Cross the river and check the Brandon quarries past Patterson Road, too. For the Rte. 6 Ponds, return to Rte. 6 and follow it west to the ponds (former quarries) on the south side of the road near Channahon.

Trails—Scope the river from the road or from under the bridge. Check above and below the dam and in the quarries.

Note—Ducks are above the dam during hunting season.

Pilcher Park (O9)

W S S F

Hickory Creek, one of the highest quality streams in the Des Plaines River watershed, forms the southern boundary of Pilcher Park. Four hundred and twenty acres of forested bottomland and ravines in the park entice a wide variety of landbirds during migration. A good number of these have also spent the summer including Cooper's Hawk, Red-shouldered Hawk, Broad-winged Hawk, Acadian Flycatcher, Brown Creeper, Veery, Wood Thrush, Black-and-white Warbler, Louisiana Waterthrush, Ovenbird, and Kentucky Warbler. Evening Grosbeaks and other winter finches have visited the feeders maintained by park staff. The naturalists offer activities for young birders.

Directions—From I-80, exit onto Briggs Street northbound. At Cass Street (Rte. 30), turn east to Highland Park Drive. This road veers left just before Rte. 30 goes under a railway overpass. Turn left on Highland Park Drive and follow the road along the creek and look for signs for the Nature Center. Parking is available in pull-outs along the drives or in designated parking areas. The park is open from dawn to dusk. The Nature Center is open from 9:00 A.M. to 4:30 P.M. on weekdays and 10:00 until 4:30 on weekends.

Trails—Four miles of trails are accessible from the Nature Center, and trail maps are available. The northern trail passes through upland woods and is favored by birders.

Owned By—Joliet Park District

Telephone—(815)741-7277

Note—Also owned by the Joliet Park District, **Higginbotham Woods** (O10) offers birding opportunities very similar to Pilcher Park, which is just to the west. From Cass Street (Rte. 30), turn east to Gougar Road. Turn north on Gougar Road to Francis Road eastbound and park in the lot off Francis Road. An old gravel road traverses the woods from east to west. Walk south through the bow-hunting area to the southeast corner.

Messenger Woods Nature Preserve (O11)

W S S F

While the assessment of beauty is certainly subjective, there is little doubt that the acres of Virginia Bluebells, Blue-eyed Mary, and trillium carpeting the swells of this preserve present a vernal aspect that is not surpassed by any other local woods. Come, then, for the flowers, but stay for the warblers, which can be thick in migration. Flycatchers, vireos, thrushes, and other migrants are also present, of course. Summering species have included Cooper's Hawk, Red-shouldered Hawk, Broad-winged Hawk, Veery, Ovenbird, and Louisiana Waterthrush.

Directions—Go north of Rte. 6 on Parker Road. Turn west on Bruce Road and park in the lot on the north side of the road.

Trails—Walk north from parking area.

Owned By—Will County Forest Preserve District

Telephone—(815)727-8700

Hickory Creek Preserve (O12)

This 1,800 acre preserve has woods, fields, creek, and ravines. Heavily used by migrants, the site is also rich in nesting species. Birders working the northwestern portion are apt to find American Woodcock displaying in early spring and Blue-winged Warblers and Yellow-breasted Chats breeding in the summer. The woods along Hickory Creek have yielded nesting Acadian Flycatchers, Cerulean Warblers, Ovenbirds, and Louisiana Waterthrushes, and the open area in the northeast part of the preserve has supported nesting Sedge Wrens. Other birds known to summer on the preserve include Cooper's Hawk, Broad-winged Hawk, White-eyed Vireo, Eastern Bluebird, and Scarlet Tanager.

Directions—Exit I-80 at Rte. 45 southbound. Turn west on Lincoln Highway (Rte. 30). Parking is available in several areas: for the trail near the creek, park in the lot for the Frankfort Swimming Pool on the north side of Rte. 30 or turn north on Wolf Road to Cleveland, turn east, and park at the end. One mile farther north, turn west from Wolf Road on Francis Road, then south on Town Line Road. Park at the railroad tracks.

Trails—From the lot at the end of Cleveland, walk north to Hickory Creek. From Town Line Road, go south across the railroad tracks to get into the preserve. A walk south of Rte. 30 west of Wolf Road connects to the **Old Plank Road Trail**.

Owned By—Will County Forest Preserve District

Telephone—(815)727-8700

Thorn Creek Nature Preserve (S1)

Thorn Creek flows through high quality woods on the eastern slope of the Valparaiso Moraine. The topographic variation provides an excellent opportunity to see how trees are distributed along a continuum of elevation and fire exposure (as it occurred historically): ravine bottoms rarely reached by fire are dominated by a very mesic forest of Sugar Maple and basswood; lower slopes are less mesic with heavy presence of Red Oaks; upper slopes which burned more often are rich with Red Oaks and White Oaks; and the uplands, the most exposed to fire, have numerous White Oaks and Bur Oaks. There are also flatwoods characterized by Swamp White Oaks growing in glacial potholes. In addition to the native forests, there are also extensive evergreen plantings which once attracted a Black-backed Woodpecker. In this 825-acre preserve, three kinds of owls (Eastern Screech, Great Horned, and Barred), Red-bellied Woodpeckers, Tufted Titmice, and Carolina Wrens are generally present year round. The feeder at the Nature Center is excellent for titmice and other seed-eating birds. Winter finches are present in some years, and Long-eared Owls and Northern Saw-whet Owls have been found in the conifers. Warbler migration is excellent and has included such species as Blue-winged, Golden-winged, and Connecticut Warblers and Yellow-breasted Chat. An outstanding assortment of birds have nested at Thorn Creek including Broad-winged Hawks, Yellow-billed Cuckoos, Willow Flycatchers, five vireos (including Bell's, Yellow-throated, and White-eyed), Veeries, Wood Thrushes, Ovenbirds, Kentucky Warblers, and Hooded Warblers.

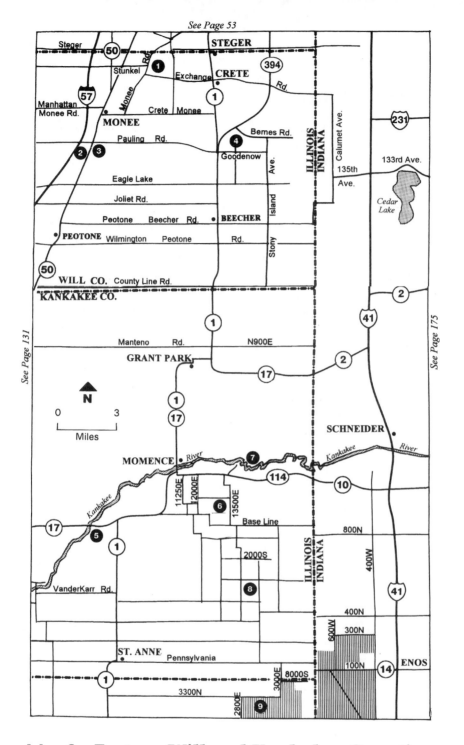

See Page 53

See Page 131

See Page 175

Map S **Eastern Will and Kankakee Counties**

Will County

1. Thorn Creek Nature Preserve *p. 141*

2. Monee Reservoir *p. 143*

3. Raccoon Grove Forest Preserve *p. 143*

4. Goodenow Grove Forest Preserve and Plum Creek Nature Center *p. 143*

Kankakee County

5. Aroma Forest Preserve *p. 145*

6. Momence Sod Farms *p. 145*

7. Momence Wetlands Nature Preserve *p. 146*

8. Pembroke Township *p. 146*

9. Iroquois County Wildlife Area *p. 147*

Directions—From I-57, exit eastbound onto Sauk Trail. Go east for one mile and turn south on Cicero (Rte. 50). At Stunkel Road (the first stop sign), turn east and proceed approximately 1.5 miles until the road ends at Monee Road. Go north on Monee to the Nature Center at 247 Monee Road (approximately .75 mile) and park in the lot on the east side of the road.

Trails—The Nature Center has displays and regularly scheduled hikes and programs. There are 2.5 miles of well-marked trails in two loops with a boardwalk on the second loop. The trail extension at the jack pines leads to small Owl Lake and cattail marsh (about 2 miles from the Nature Center).

Note—The preserve is open from 8:00 A.M. to 10:00 P.M. The Nature Center is open year round from 12:00 noon to 4:00 Thursday–Sunday.

Owned By—Multiple agencies; managed by the Thorn Creek Nature Preserve Commission

Telephone—(708)747-6320

Raccoon Grove (S3)

Rock Creek flows through this 210-acre forest, the southern half of which is a dedicated nature preserve. Red-headed Woodpeckers and Tufted Titmice are generally present throughout the year, and warblers and other passerines can be common during migration. The fields east of Egyptian Trail Road and east of the preserve have attracted wintering Short-eared Owls and summering Eastern Bluebirds and Grasshopper Sparrows.

Directions—Park in the lot on the south side of Pauling east of Monee Reservoir and Rte. 50. For the area south of the creek, park on Egyptian Trail Road (Central Avenue) on the east side of the preserve.

Trails—There is a short path from the parking lot. Rock Creek divides the preserve into a northern and a larger, southern portion. The southern portion does not have marked trails, but there is a bridle path which enters the preserve from Egyptian Trail Road.

Note—**Monee Reservoir (S2)**, a 46-acre fishing lake with a 1.5-mile hiking trail is west of Raccoon Grove. Heavy boating ruins the birding here except in late winter and early spring when various species of waterfowl pause on their way north. Park in the lot on Ridgeland Avenue south of Pauling Road.

Owned By—Will County Forest Preserve District

Telephone—(815)727-8700

Plum Creek Nature Center (S4)

The Nature Center is located in **Goodenow Grove**, a 689-acre preserve consisting of woods, fields, and a kame. Winter birding is best at the Nature Center feeders (Tufted Titmice, American Tree Sparrows, Red-breasted Nuthatches, and Purple Finches) and the fields to the south of the preserve (Horned Larks, Lapland

Longspurs, and Snow Buntings). Spring and fall migration brings large numbers of flycatchers, vireos, thrushes, and warblers. An unusually varied group of birds have summered here including Broad-winged Hawk, American Woodcock, Barred Owl, Red-headed Woodpecker, Eastern Phoebe, White-eyed Vireo, Bell's Vireo, Yellow-breasted Chat, Scarlet Tanager, and Orchard Oriole. Western Meadowlarks are regular summer residents in the fields east and south of the towns of Beecher and Peotone.

Directions—Go east from the junction of Highway 1 and Rte. 394 on Goodenow Road. Look for the sign for the preserve and Nature Center on the north side of the road at Dutton Road and turn north to the preserve. The preserve opens at 8:00 A.M.; the Nature Center is open from 10:00 A.M. to 4:00 P.M.

Trails—Approximately 3 miles of trails begin at the Nature Center and pass through open areas and woods.

Owned By—Will County Forest Preserve District

Telephone—(708)946-2216

Kankakee County

Kankakee River State Park (R10)

W S S F

This mostly wooded 4,000-acre state park protects both sides of the river as well as several islands. The state has dedicated 139 acres as a nature preserve. Undoubtedly, the parks's most remarkable organism is the Kankakee Mallow, a Hollyhock look-alike that grows naturally nowhere else in the world. Most birders visit in spring and fall when migrating landbirds are often abundant. This area also makes a rewarding summer destination, for there is a fine variety of nesting species: Turkey Vulture, Acadian Flycatcher, Cerulean Warbler, American Redstart, Prothonotary Warbler, Ovenbird, Kentucky Warbler, and Scarlet Tanager.

Directions—Drive northwest from Kankakee on Rte. 102 for the north side of the park or on Rte. 113 for the south side of the park. The nature preserve is on both sides of the river in the western part of the park, and includes the island known as Altorf or Langham Island in the eastern part of the park. The visitors' center and Nature Center are on Rte. 102.

Trails—This is a long, narrow preserve. Park in one of the lots along either highway to bird, then move along and park in the next lot.

Owned By—Illinois Department of Natural Resources

Telephone—(815)933-1383

Note—Two preserves south of Kankakee may be worth a visit. **Gar Creek Trail and Prairie** (R11) is south of Kankakee on River Road east of Highway 45/52, and **Iroquois Woods Natural Area** (R12) is south of Aroma Park on County Road 5000S.

Aroma Forest Preserve (S5)

Lying on the east bank of the Kankakee River, the 55 acres of evergreens, dry oak woods, sand prairie, and bottomland woods that comprise this preserve often produce good numbers of migrant passerines and has an interesting selection of breeding species. Among those known to nest here are Tufted Titmice, Wood Thrushes, and Prothonotary Warblers. The Kankakee County Audubon Society schedules regular field trips to the preserve.

Directions—Drive south on I-57 and turn east on Rte. 17 in Kankakee. Drive east for approximately 5 miles and turn south on Heiland Road after crossing the Kankakee River. The Forest Preserve is 1.4 miles south on Heiland Road.

Trails—A 1-mile trail heads west from the lot behind wooded backyards, turns north, and passes through the prairie, dry woods, and bottomland woods and ends at the river.

Owned By—Kankakee River Valley Forest Preserve District

Telephone—(815)935-5630

Momence Sod Farms (S6)

Acres of fields planted in sod and other crops, including colorful and aromatic garlic, attract a host of shorebirds and other species. Landbirds known to nest in the area include Northern Mockingbird, Vesper Sparrow, Grasshopper Sparrow, Dickcissel, and Western Meadowlark. As shorebird habitat declines throughout the region, an increasing number of birders make the drive to Momence to see the parade of species that begins with Upland Sandpipers in mid-July. Then about a month later, the Uplands are joined by Black-bellied Plovers, American Golden-Plovers, Pectoral Sandpipers, and Buff-breasted Sandpipers. Concentrations are highest immediately following the watering of the sod. Small flocks of Semipalmated and Least Sandpipers also arrive in mid-August, along with a few Baird's Sandpipers in their midst. If there is standing water, look for dowitchers, both Yellowlegs, Solitary and Stilt Sandpipers, Common Snipe, and Wilson's Phalaropes (uncommon). Rarer species have included Hudsonian Godwit, Ruddy Turnstones, Red Knots, and Red-necked Phalarope. Most of the shorebirds depart by Labor Day, but the American Golden-Plovers linger on into October.

Directions—From Momence, drive east on Highway 114 for three miles. Turn south on County Road 13400E after passing the sign for H&E Sod Nursery on the south side of the road and the Mug and Suds Bar on the north side of the road. Many roads in this area are gravel. Bird from the road and look for sod fields that have been recently watered or stripped of sod leaving moist soil exposed. The location of the concentrations of shorebirds varies from year to year. Follow 13400E south as it bends east, then south again on 13500E. Pass 1250N and look for Upland Sandpipers on the west side of the road. Return to 1250N and turn west. Continue to the end of 1250N, turn south on 12000E, then west on 1000N. Be careful here as the traffic increases on the paved roads. At 11250E turn north and look for the sod fields on the east side of the road. This is where a Whimbrel and Marbled Godwit were seen one August. Follow County Road

11250E north to Highway 114 and back to Momence. Respect the property owners and bird from the road.

Momence Wetlands Nature Preserve (S7)

W S S F

From its origins near South Bend, Indiana, to Momence, Illinois, where a limestone ledge impeded its flow, the Kankakee River used to be a slow-moving river with 2,000 bends. By the early part of this century, the part of the river in Indiana had been converted into a series of ditches. But at the Illinois line, the river resumed its winding course through marsh and forest. This section of the river is known as the Momence Wetlands, a haven for wildlife that remains accessible only by small boat. Barred Owls, Pileated Woodpeckers, and Tufted Titmice are permanent residents. During migration, vireos, thrushes, and warblers move through in significant numbers. Species recorded here in summer include Hooded Merganser, Red-shouldered Hawk, Carolina Wren, Veery, American Redstart, Prothonotary Warbler, Louisiana Waterthrush, and Scarlet Tanager.

Directions—No trails or parking lots exist. Visiting the area by canoe or kayak may be the best way to see the area. To approach the area, go through Momence and turn east on Highway 114. Turn north in about three miles after passing the Mug and Suds bar onto County Road 13350E. This road passes through wet woods and ends at a turnaround opposite the western edge of the nature preserve. A canoe can be launched here to explore the oxbows and sloughs to the east. A smaller section of the preserve is further east on the west side of County Road 17000E, but no access is available at this time. State Line Road to the east provides access to similar habitat.

Owned By—Illinois Department of Natural Resources

Pembroke Township (S8)

W S S F

Frequent burning has maintained the black oak savannas of Pembroke Township. Birds that nest in the area include Northern Bobwhite, Eastern Screech-Owl, Whip-poor-will, Red-headed Woodpecker, Bell's Vireo, Northern Mockingbird, Yellow-breasted Chat, Grasshopper Sparrow, and Orchard Oriole. Carolina Wrens and Lark Sparrows nest here too but are less common than the other species.

Directions—From Momence, go east on Highway 114 and turn south on County Road 11250E, as soon as you cross the railroad tracks. This road passes an area of sod farms (described above). It makes several right-angle turns, and after about eight such bends in the road, you come to County Road 2000S. Turn east here and proceed one mile to the intersection with 14000E. Bell's Vireos like the willows on the southeast corner of this intersection. Orchard Orioles have nested in the tall trees on the northeastern corner, and Yellow-breasted Chats have nested in the area, too. From this intersection, drive north on 14000E, jog west to 13750E, and turn north. In this area, a Blue Grosbeak was observed singing in July of 1997. Drive other roads in the area slowly and listen for birds.

Many of the roads in this area are unpaved, and the sandy soil is quite soft. The nearby community is Hopkins Park, and the residents here do not seem to be bothered by birders, but small packs of dogs may come up to your vehicle to sniff and bark at your door.

Note—South of the Pembroke area and the Kankakee County line, the 2,000-acre **Iroquois County Conservation Area** (S9) has excellent habitat for birds. This area of wet meadows and marshes allows public access during non-hunting seasons. The adjacent Hooper Branch Savanna Nature Preserve is wooded with Pin Oaks and Black Oaks. Special birds of the area include Yellow-breasted Chats, Summer Tanagers, Vesper Sparrows, Lark Sparrows, and Henslow's Sparrows. To reach the area, drive south to County Road 3300N in Iroquois County. The headquarters is located at the intersection of roads 3300N and 2800E, and the Nature Preserve is located east of the intersection of roads 3300N and 3000E.

Owned By—Illinois Department of Natural Resources

Telephone—(815)435-2218

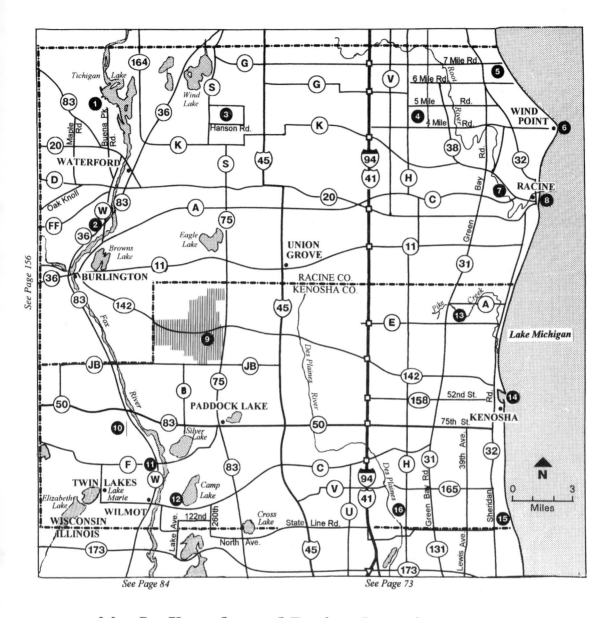

See Page 156

See Page 84

See Page 73

Map B **Kenosha and Racine Counties**

Racine County

1. Tichigan Marsh *p. 152*
2. Rochester Area *p. 152*
3. Wind Lake Sod Farms *p. 153*
4. Nicholson Pond *p. 153*
5. Cliffside Park *p. 153*
6. Wind Point *p. 154*
7. Colonial Park *p. 154*
8. Racine Lakefront *p. 155*

Kenosha County

9. Bong State Recreation Area *p. 149*
10. New Munster Wildlife Area *p. 150*
11. Fox River Park *p. 150*
12. Camp Lake *p. 150*
13. Petrifying Springs Park *p. 150*
14. Kenosha Harbor *p. 151*
15. Chiwaukee Prairie *p. 151*
16. Prairie Spring Park *p. 151*

Kenosha County

Bong State Recreation Area (B9)

W S S F

In 1958 the U.S. government began construction of a 4,515-acre jet fighter/interceptor base. Contractors removed top-soil, filled wetlands, and began to lay the foundation of runways with more than a million cubic yards of gravel. The project was abandoned one year later, and the land remained idle until it was acquired by the Wisconsin Department of Natural Resources in 1974. Today, Bong State Recreation Area's fields and marshes are used primarily for such recreational activities as hunting, snowmobiling, motorized biking, hang gliding, horseback riding, camping, and nature study. Birders find it to be a good location for wintering hawks, Lapland Longspurs, and Snow Buntings, but the star attractions are the Short-eared Owls that are best found from December through February. During the day, the owls hunker down in the shelter of south-facing gullies, waiting to emerge at dusk to hunt small mammals in the grasslands south of the park headquarters. Other owl species frequently roost in the pines on the north side of Highway 142, approximately .25 mile east of the entrance. Early spring brings flocks of migrants to Wolf Lake, as well as the smaller ponds and marshes in the area. Common Loons, Horned Grebes, American White Pelicans, Greater White-fronted Geese, Snow Geese, Ospreys, and Bald Eagles are among the birds to look for. Warblers, arriving later, prefer the woods on the north side of Highway 142. The marshes and fields of Bong have produced an impressive list of nesting species: Least Bittern, King Rail, Virginia Rail, Sora, Upland Sandpiper, Forster's Tern, Black Tern, Horned Lark, Sedge Wren, eight sparrows (including Vesper, Grasshopper, and Henslow's), Bobolink, Yellow-headed Blackbird, and Orchard Oriole (near the park headquarters).

Directions—Drive west on Highway 142. The entrance to the State Recreation Area is on the south side of Highway 142 west of Highway 75. An entrance fee is charged. Park in the lot for the visitors' center to pick up a map and ask the naturalist about recent sightings of birds. Roads lead through open fields to several other parking areas where marsh and open water areas may be viewed. Farther west, the grove of cottonwoods at the Vista picnic area off of County Highway "B" north of Highway 142 is a good place for warblers in migration. A daily entrance fee is charged to park ($7 in 1999, or $25 for an annual permit)

Owned By—Wisconsin Department of Natural Resources

Telephone—(262)878-5600

New Munster Wildlife Area (B10)

W S S F

This 1,009-acre wildlife area is comprised of a large marsh surrounded by woods, pine plantations, and open fields. Hunting is allowed in season. During May and October, brushy hedgerows often yield large flocks of sparrows, which have occasionally contained Harris's Sparrows. Nesting species have included such upland birds as Blue-winged Warbler and Yellow-breasted Chat, and such marsh birds as Least Bittern, Virginia Rail, Sora, Sedge Wren, and Marsh Wren.

Directions—Drive south from Highway 50 on Highway "W" and Highway "JI." At 73rd Street, turn west. This road becomes Lily Lake Road. Park at the south end of Lily Lake Road. The marsh is west and south of this point.

Owned By—Wisconsin Department of Natural Resources

Telephone—(262)878-5600

Camp Lake (B12)

W S S F

This inland lake with extensive cattail marshes is good for migrant waterfowl and nesting King Rails, Virginia Rails, and Soras.

Directions—Drive three miles west of Highway 83 on County Highway "C." Park on Highway "C" and bird from the road or turn north on the first road west of the marsh.

Note—Nearby Fox River has a wooded county park, **Fox River Park** (B11), and an area of backwaters located north of the park off of County Road "W."

Petrifying Springs Park (B13)

W S S F

Winter is the best season to bird these 360 rolling acres along Pike Creek. The extensive conifer plantings have provided roosting space for up to six species of owls: Eastern Screech, Great Horned, Barred, Long-eared, Short-eared, and Northern Saw-whet. (A seventh species, Snowy Owl, has also been seen at the park.) The conifer tract at the northeastern corner of Highway 31 and Highway "E" and another near the baseball diamond are among the most productive in the park. Other winter birds have included Sharp-shinned Hawk, Cooper's Hawk, Northern Goshawk, and Rough-legged Hawk. The deciduous woods along the creek are also worth birding for migrant passerines.

Directions—Drive east of I-94 on Highway "E" to Highway 31. Drive north on 31 (Green Bay Road) to Highway "JR," then east .5 mile to parking lots for the cross country course and the entrance to Petrifying Springs Park. The park can also be entered from the north off of Highway "A." The 40-acre **Hawthorn Hollow Nature Sanctuary** is adjacent (see directions in Appendix B: Nature Centers).

Trails—North of County "A," a trail on top of a ravine parallels Pike Creek. The site can be birded in winter on cross country skis.

Owned By—Kenosha County Parks

Telephone—(262)857-1869

Kenosha Harbor (B14)

The Kenosha Harbor complex, with its jetties, impoundment, and beaches, presents the best birding on the Kenosha lakefront. From November through March, Greater Scaup, White-winged Scoters, Oldsquaws, Common Goldeneyes, mergansers, and other ducks favoring Lake Michigan raft within the harbor or the less-protected harbor mouth. Scan the flocks within the harbor for Horned Grebes, Gadwall, American Wigeon, and other waterfowl. Snowy Owls can be anywhere in the vicinity, while winter gulls (including Iceland and Glaucous) prefer the areas near the Holiday Inn. During migration, the breakwater and the impoundment on the south side of the harbor are good places to look for herons, shorebirds, gulls, and terns.

Directions—Drive east from I-94 on Highway 158 to 6th Avenue and the harbor. Turn north on 6th Avenue and view the harbor from behind the Holiday Inn. Continue north on 6th Avenue to 50th Street and turn east toward Simmons Island Park for a view of the entrance to the harbor and breakwater. Use a telescope to check the breakwater and the impoundment on the south side of the entrance to the harbor. To get closer to the south pier and to see the parks south of the harbor, turn east from 6th Avenue on 57th Street toward the Southport Marina and drive past the boat storage area toward the lighthouse. Another park, **Southport Park**, is farther south at 78th Street and Lake Michigan next to Kenosha's water treatment plant. This is a good location for gulls in winter.

Chiwaukee Prairie (B15)

This 165 acres of Chicago Lake Plain prairie is one of the region's outstanding treasures. A visit here in late May is unforgettable: acres of Shooting Stars, ranging from deep amethyst to white, mark the gentle ridges that parallel the lake shore. Birds-foot Violet, pucoon, Golden Alexander, and Wood Betony contribute additional hues to the scene. Merlins, Peregrine Falcons, Short-eared Owls, sparrows (including Le Conte's and Nelson's Sharp-tailed), and other lakefront migrants use the site, as do nesting Least Bitterns, Virginia Rails, Upland Sandpipers, Common Snipe, Swamp Sparrows, Bobolinks, Orchard Orioles, and other species.

Directions—From Sheridan Road, turn east on 116th Street. Chiwaukee Prairie is on the south side of 116th Street after crossing the railroad tracks. Look for the roads to the south and park where indicated.

Trails—Walk on established paths to avoid disturbing plants and nesting birds.

Prairie Spring Park (B16)

The centerpiece of this park is Andrea Lake, forty feet deep and spring fed. In spring and fall it is frequented by Common Loons and numerous ducks. Surrounding the lake are fields, woods, and marshes that provide habitat for a variety of nesting and migrant species.

Directions—Turn east from I-94 on 104th Street and proceed to Highway "H" in Pleasant Prairie. Turn north to the entrance.

Trails—The lake is easily visible from the road and numerous trails lead into the upland areas.

Owned By—Town of Pleasant Prairie

Racine County

Tichigan Marsh (B1)

W S S F

This 1,221-acre hunting area of marsh, fields, and brushy areas on the shore of Tichigan Lake is managed principally for migrant waterfowl. When the water is drawn down, it often attracts shorebirds. Nesting species include Sandhill Crane, Forster's Tern, Willow Flycatcher, Sedge Wren, and Marsh Wren.

Directions—Travel west of Waterford on Highway 20/83. At Buena Park Road, turn north for 2.75 miles to the first parking area. Another 1.8 miles north will bring you to Bridge Road. Turn east .6 mile to the parking area for Tichigan Lake. The impounded area is .5 mile south of this parking area, although it is generally closed after August 31.

Owned By—Wisconsin Department of Natural Resources

Telephone—(262)878-5600

Rochester Area (B2)

W S S F

Four birding areas with a diverse range of habitats, flank the Fox River between Highway 20 on the north and Highway 11 on the south. Elas Woods (also known as Case Eagle Woods) and Wehmhoff Woodland Preserve are very good for migrants and have had nesting Ruby-throated Hummingbirds, Blue-gray Gnatcatchers, Wood Thrushes, Blue-winged Warblers, Chestnut-sided Warblers, and Scarlet Tanagers. A third site, on the west side of the river, is called Rochester Public Hunting Ground, and it contains a tamarack bog. Good during spring and fall, the bog has also held summering Alder Flycatchers, Northern Waterthrushes, and Mourning Warblers. Lastly, Saller Woods is an area of mixed deciduous trees and evergreens that is interesting for migrant land birds and wintering owls and finches.

Directions—For the 239-acre Elas Woods area on the east side of the river, turn south on County Highway "J" from County Highway "D." Bear right on South River Road for .2 mile and park near the sign for the Snowmobile Trails of Racine County Parks. Follow the dirt road west into Elas Woods. For the bog on the west side of the river, travel south on County Highway "W" from County Highway "D." Park in the lot on the west side of the road after one mile to bird the large tract which includes a bog. For the Wehmhoff Woodland Preserve, travel west on County Highway "FF" from County Highway "W." Turn left on South Honey Lake Road and park in the lot after .5 mile. A second lot for the preserve is another .4 mile south. The lot for Saller Woods is off of Rte. 36, just west of the Fox River.

Wind Lake Sod Farms (B3)

From late July through the end of September, an impressive suite of shorebirds flock to this amalgam of sod fields, irrigation ditches, canals, and Wind Lake. Birds seen have included American Golden-Plover, Black-bellied Plover, Upland Sandpiper (late July and early August), Least Sandpiper, Baird's Sandpiper, Pectoral Sandpiper, and Short-billed Dowitcher. The specialty of the site, however, is the Buff-breasted Sandpiper, which can be expected on the sod fields from the last week of August though the second week of September. But the area is not for shorebirds alone. An abundance of dabbling and diving ducks collect at Wind Lake in early spring, Lapland Longspurs linger in spring and fall, and Brewer's Blackbirds nest along the irrigation ditches throughout the sod farm area. In the fall, all six swallows are common, and Horned Larks are abundant.

Directions—Drive north from Highway 20 on Highway 45. At County Highway "G," turn west through the first sod farms and turn south on County "S." For a view of Wind Lake, go west on Wind Lake Road and south on West Loomis Road to the lake. The best view is from Sportsmen's Boat Rental. For more sod farms, retrace your path to County "S" and turn south. Go east on Burmeister Road, south on Britton Road, then west on Hansen Road to return to County "S."

Owned By—private owners

Nicholson Pond (B4)

This 200-acre wildlife preserve consists of a creek, an evergreen plantation, and fields that usually flood in spring. In the winter, Long-eared Owls sometimes roost in the evergreens, and Horned Larks, Lapland Longspurs, and Snow Buntings forage in the fields. But the best birding is in the spring when the fields are flooded and hundreds of geese, ducks, shorebirds, and other species stop on their way north. Look for the same mix in late summer and fall if the ponds persist. Gray Partridge have also been seen here.

Directions—Drive .5 mile east of County "H" on Five Mile Road. Look for the pine plantation on the south side of the road and park in the small lot at the pines. Another vantage point is from Four Mile Road on the south, but there is no off-road parking.

Trails—Enter from the Louise Erickson Nature Trail on 5 Mile Road.

Cliffside Park (B5)

An excellent example of the Lake Border Upland with eroding bluffs and a ravine harboring rich mesic forest, the 232-acre recreational park also contains extensive fields. Because of its lakefront location and diverse habitat, Cliffside Park is an excellent location to observe migrants: water birds moving along the shore and warblers and other land birds in the woods and fields. It is the county's best site for the fall hawk migration. Upland Sandpipers, Grasshopper Sparrows, and Bobolinks have nested in the fields.

Directions—Travel east on Six Mile Road (west of Rte. 32 the road is called County Road "G"). After crossing the railroad tracks, turn left immediately on Michna Road to the park entrance. Lake Michigan is not visible from the parking lot.

Trails—For the shortest walk to the lake and fields, park in picnic area #2, and for the shortest walk to the ravine, park in area #3.

Owned By—Racine County Parks

Telephone—(262)888-8440

Wind Point (B6)

W S S F

Extending into Lake Michigan, Wind Point is a prime location to observe the lakefront migration of birds, particularly during inclement weather. A remarkable example is the 18,000 Common Nighthawks that flew over in one hour during a day in late August. In addition to common species, an impressive list of unusual birds have been seen from this small spit of land. Red-throated Loon, Red-necked Grebe, Harlequin Duck, all three scoters, Piping Plover, Whimbrel, Red Knot, Western Sandpiper, Purple Sandpiper, Red-necked Phalarope, jaegers, Laughing Gull, and Little Gull all occur here on occasion.

Directions—Drive east on Four Mile Road from Highway 31 until the road turns right and becomes Lighthouse Drive. This road will lead to Shoop Park and the lighthouse. Park at the lighthouse or in the lot for the golf course to the south.

Trails—The lake and shoreline are accessible from either parking lot. If golfers can be avoided, search the wooded edges of the golf course and check the ravine at the south end of the park.

Note—The adjoining property is owned by the Wingspread Foundation and protects the 35-acre estate of H. F. Johnson and his home designed by Frank Lloyd Wright. Conferences are held here, but the estate and grounds are open to the public on weekdays from 9:00 A.M. to 4:00 P.M. if no conferences are being held. Call (262)681-3353 to learn if a visit is possible.

Colonial Park (B7)

W S S F

These lowland woods along the Root River may be the county's best spot for migrant warblers and other passerines, particularly in May. The warbler list includes such rarities as Prothonotary, Kentucky, Hooded, and even one Kirtland's. Green Herons, Wood Ducks, and Spotted Sandpipers are present in the summer.

Directions—Travel 2.1 miles east of Highway 31 on Highway 38. At West High Street, follow the sign to the park.

Trails—Walk on the trails on either side of the river.

Owned By—City of Racine

Telephone—(262)636-9131

Racine Lakefront (B8)

During spring and fall, birders perched on the observation platform overlooking the lake are apt to see loons, grebes, Double-crested Cormorants, gulls, and terns. The boat launch, North Beach, and the small beach south of the harbor all attract shorebirds, gulls, and terns. Fifteen to twenty-five Canvasbacks usually spend the winter in the breakwater area.

Directions—Drive toward the lake on Rte. 20 (Washington Avenue). Continue past Main Street on 6th Street to the boat launch on the harbor. For another view of this area and a look at Lake Michigan, drive east on 4th Street. Pass the yacht club and restaurant to the platform which overlooks Lake Michigan. To view the northern part of the harbor, drive north on Main Street over the Root River bridge and turn right immediately onto Dodge Street. Continue east; jog north on Michigan Street and immediately east on Reichert Court. The harbor is visible at the end of the road. To view the lake north of the harbor, continue north on Michigan Street for a few blocks, then turn east to view North Beach. South of the harbor, a small beach and the protected waters inside the breakwater can be seen by driving south on Main Street to Sam Myers Park on the lake at 11th Street or to the bluff on 16th Street.

Owned By—City of Racine

Telephone—(262)636-9131

Walworth County

Lake Geneva (A7)

With the exception of Lake Michigan, Lake Geneva is the largest and deepest lake in the region. As a result, it remains ice-free longer than other inland lakes. The grebes, Common Goldeneyes, and Common Mergansers that arrive in fall stay until the water freezes over (later than in other inland lakes) and appear again in the spring when the ice breaks up. Waterfowl numbers are often impressive. Lake Geneva fills a large morainal basin whose wooded slopes have such nesting species as Cooper's Hawk, Red-headed Woodpecker, Ruby-throated Hummingbird, Yellow-throated Vireo, and Scarlet Tanager. Black Terns can also be seen in the summer as they search for food over the lake.

Directions—The lake is visible from Highway 50 in Lake Geneva, Rte. 120 on the eastern shore, Linn Pier on the south shore, and Rte. 67 in Fontana and Williams Bay.

Trails—A 26-mile footpath follows the shoreline around the lake. Park in any of the towns or at any of the boat ramps to access the path.

Note—Several wetland areas east of Lake Geneva attract nesting Sandhill Cranes, Black Terns, Yellow-headed Blackbirds, and migrating waterfowl and passerines. **Townline Pond** (A12) is on Town Line Road west of South Road; **Ivanhoe Marsh State Wildlife Area** (A14) is on the south side of Bloomfield Road three miles

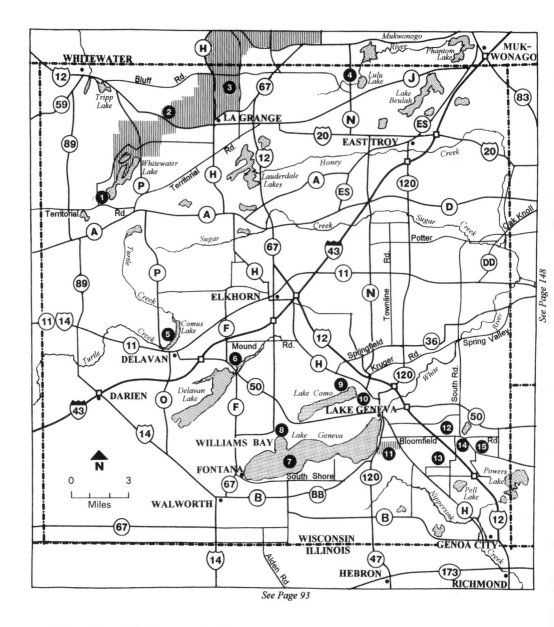

See Page 93

Map A **Walworth County**

1. Natureland County Park *p. 159*
2. Lake La Grange *p. 158*
3. Kettle Moraine State Forest *p. 158*
4. Lulu Lake *p. 159*
5. Comus Lake *p. 158*
6. Delavan Inlet *p. 158*
7. Lake Geneva *p. 155*
8. Kishwauketoe Nature Preserve *p. 157*
9. Lake Como *p. 157*
10. Duck Lake Nature Area *p. 158*
11. Big Foot Beach State Park *p. 157*
12. Townline Pond *p. 155*
13. Bloomfield State Wildlife Area *p. 157*
14. Ivanhoe Marsh State Wildlife Area *p. 155*
15. Haf's Pond *p. 157*

east of Lake Geneva; **Haf's Pond** (A15) is further east on Haf Road off Bloomfield Road; Four Seasons Nature Center is on Hightway "H"; and **Bloomfield State Wildlife Area** (A13) is on Litchfield Road east of Highway "H."

Big Foot Beach State Park (A11)

W S S F

This 272-acre park is located at the east end of Lake Geneva. The hills are covered with oaks, the low areas around the lagoon near Lake Geneva have many European alders, and the northeast part of the park behind the high school has an evergreen plantation. The southern portion has fields and marsh. In winter, the park is a good place to look for owls, Rusty Blackbirds, and Common Redpolls (in the alders). Spring and fall can also be productive for migrant passerines.

Directions—Drive south from Highway 50 in Lake Geneva on Highway 120 to the main entrance of the park (closed in winter). The winter entrance is on County Highway "H" south of the high school and Bloomfield Road (the first drive south of the pines). This entrance is difficult to locate as there is no sign posted. An entrance fee or annual sticker is required.

Trails—Five miles of hiking trails and a .5 mile self-guided nature trail exist. The trails are easy to use in winter for cross country skiing. In summer, the park is very crowded, especially on weekends.

Owned By—Wisconsin Department of Natural Resources

Telephone—(262)248-2528

Kishwauketoe Nature Preserve (A8)

W S S F

The wetlands, prairie, and upland woods included within this 231-acre preserve attract numerous migrants and such nesting species as Cooper's Hawk, Willow Flycatcher, Eastern Bluebird, and Blue-winged Warbler.

Directions—Turn south of Highway 50 on Rte. 67 and park in the Lion's Club Fieldhouse on the west side of Highway 67.

Trails—Four miles of trails and a boardwalk traverse the area. A guide and map are available at the trailhead.

Owned By—Town of Williams Bay

Lake Como (A9)

W S S F

This shallow lake is three miles long and has extensive cattail marshes at both ends. Numerous water birds, including thousands of American Coot, gather here in migration, and Black Terns, Sedge Wrens, Yellow Warblers, and Swamp Sparrows nest in the marshy edges.

Directions—The lake can be viewed from several spots. For the southern shore, turn north from Highway 50 on Schofield Road two miles west of Lake Geneva.

For the eastern shore, travel 1.5 miles north from Lake Geneva on County Highway "H." After crossing Como Creek, look for the small gravel road on the west side of the highway leading to a small parking area. Views can be had from the town of Como on the north shore.

Trails—View the lake from the shore or launch a canoe from either parking area.

Note—**Duck Lake Nature Area** (A10), and the two-mile Duck Lake Trail (also known as **Warbler Walkway**) are on the south shore of the lake where woods and brush provide excellent habitat during migration for flycatchers, vireos, thrushes, warblers, and sparrows. White-eyed Vireos and Yellow-breasted Chats are regularly seen, and Yellow Warblers and Swamp Sparrows are among the nesting species.

Directions—Park at the lot off Highway "H" for the east end of the area or near the boat ramp on Schofield Road. The trail begins south of the dam near Highway "H" and continues west to Schofield Road.

Delavan Inlet (A6)

W S S F

Many migrant ducks populate the shallow inlet and creek on the north end of Delavan Lake. The marshes and open water also accommodate summering Hooded Mergansers, Sandhill Cranes, Forster's Terns, Black Terns, and Yellow-headed Blackbirds.

Directions—Park at Reeds Marine (which is privately owned) on Highway 50 west of the inlet, or drive north on Inlet Shore Drive just west of the inlet and watch for the gravel road which ends at the inlet opposite Delavan Drive. Another vantage point is at the north end on Mound Road Between County Highway "F" and Rte. 67.

Trails—Bird from the parking lot or road. Canoes can be launched from any of the locations for the best view of the area.

Note—Another shallow lake, **Comus Lake** (A5), is north of Delavan between 7th Street and Dam Road.

Kettle Moraine State Forest (A3)

W S S F

Some of the world's finest examples of kettles, kames, eskers, drumlins, and other glacial landforms are preserved in the Southern and Northern Units of the Kettle Moraine State Forest. All of the Northern Unit and much of the of the Southern Unit is north of the Chicago Region, but the portion in Walworth County does provide good birding during the periods of migration and breeding. Uncommon nesting species include Least Bittern, both cuckoos, Whip-poor-wills, Sedge Wren, Blue-winged Warbler, Golden-winged Warbler, Western Meadowlark, and Orchard Oriole. One particular location worth checking is **Lake La Grange** (A2), a 500-acre area of oak, hickory, and aspen woods on a moraine overlooking the small lake. It can be excellent for warblers in migration. Yellow-throated Vireos and Scarlet Tanagers nest here, and the lake can have herons when the water levels are low. Wild Turkeys are sometimes seen along roads in this area.

Directions—For Lake La Grange, travel 2.8 miles west of the intersection of Rte. 12 and County Highway "H" in La Grange. Park in the gravel lot on the north side of the road. Alternatively, you can drive the gravel roads to the east and north of this point and park along the roads. For park headquarters, go north of Walworth County on Rte. 67 to the town of Eagle and turn west on Highway 59. The visitors' center and a small nature trail are on the south side of Highway 59, and maps are available.

Trails—Purchase a map of the Kettle Moraine State Forest in the La Grange General Store. A good sandwich can be purchased here, too. Walk the gravel roads. The Ice Age Trail traverses this area, and horseback riders use the trails.

Owned By—Wisconsin Department of Natural Resources

Telephone—(262)594-2135

Note—**Natureland County Park** (A1) is located nearby on the south shore of Whitewater Lake.

Lulu Lake (A4)

Although this 350-acre preserve is open only for Nature Conservancy tours, a good view of its sedge meadows, stream and marsh can be had from the roadside. Sandhill Cranes nest in the area and can be heard from Nature Road. Alder Flycatchers have also nested here.

W S S F

Directions—Turn north from Rte. 20 on County "N." Cross County "J" where the road becomes Nature Road. After one mile, pause at the stream crossing to listen for cranes and proceed north another mile to observe the marsh.

Trails—Bird from Nature Road. Entrance to the Nature Conservancy property is from N9564 Nature Road.

Note—**Beulah Bog**, a small but diverse preserve with a trail leading to a kettle depression bog, is located two miles to the east. The different plant communities support a rich mix of birds, including summering Alder Flycatcher and Hooded and Nashville Warblers. Owned by the Department of Natural Resources, the preserve is on the east side of Stringers Bridge Road approximately one mile south of County "J."

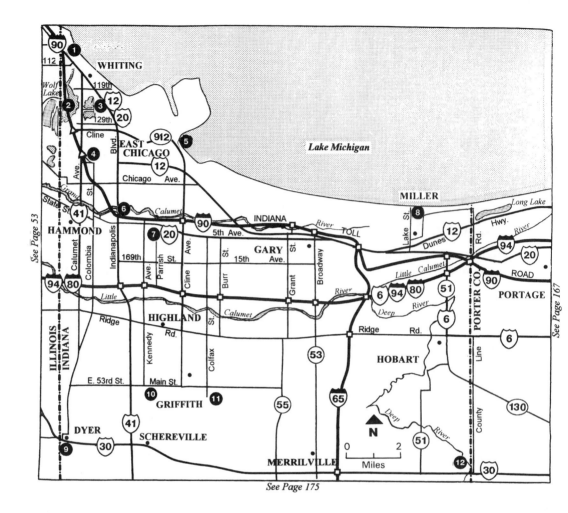

Map V **Northwestern Indiana**

1. Migrant Trap *p. 161*

2. Wolf Lake *p. 162*

3. Lake George *p. 162*

4. Hammond Cinder Flats *p. 162*

5. Jeorse Park *p. 162*

6. Roxana Pond *p. 163*

7. Gibson Woods Nature Preserve *p. 163*

8. Miller Beach *p. 164*

9. St. Margaret Mercy Health Care Center *p. 65*

10. Hoosier Prairie Nature Preserve *p. 165*

11. Oak Ridge Prairie County Park *p. 164*

12. Deep River County Park *p. 165*

Lake County

Hammond Lakefront Park and Sanctuary ("The Migrant Trap") (V1)

According to Ken Brock (1997, 11), this site, comprised of 9.5 acres of fill, "is almost certainly the best single location in the region for observing migrant passerines." Over the years, blowing sand has partially covered chunks of concrete originally stockpiled to build a breakwater. Cottonwoods and other early successional vegetation have taken root in the nooks and crannies of this artificial dune, thus providing the only cover in an otherwise heavily industrialized section of lakefront. While excellent birding can be had in May, the period from mid-August through October is generally most productive, as southbound migrants seek refuge in this vegetated sanctuary. Brock cites the remarkable fact that one-quarter of all recent Connecticut Warbler records from the Indiana Dunes region come from this tiny site. Virginia Rails, Soras, American Woodcocks, Short-eared Owls, and Whip-poor-wills are among the non-passerines that have also put in appearances. Winter birding has yielded White-winged Scoters, Glaucous and Great Black-backed Gulls, and Snowy Owls. It is also worth mentioning that, as the frequent visitor will notice, birds aren't the only flying organisms that migrate along the lake. Red Bats, Silver-haired Bats, and Monarch Butterflies move south from late August through September. Perhaps the largest concentration of Monarchs ever recorded in the Chicago region occurred here one morning in September 1989 when two observers estimated the presence of over 100,000 butterflies.

W S S F

Directions—From Indianapolis Boulevard, take the overpass to the Hammond Marina and Empress Casino on Lake Michigan. Turn off at the sign for overflow parking and the marina and park near the marina gatehouse.

Trails—Walk through the gate in the fence toward the lake. Several trails parallel the shoreline, but all are very rough. Be sure to check the leeward side of the ridge for birds seeking shelter from the winds.

Owned By—Hammond Parks Foundation

Note—About 1.25 miles to the east is **Whiting Park**, a lakefront area of beach, mowed lawn, scattered trees, conifers, and weedy area. This public park is less isolated and better manicured than the migrant trap, so the birding is less productive, although possibly more comfortable. Still, the site attracts its share of migrants, particularly sparrows (Vesper, Savannah, Le Conte's, and Harris's) in October. To get to Whiting Park, go east of Indianapolis Boulevard on 119th Street in Whiting. This road leads directly into Whiting Park. By following the park drive around to the north, you come to **Whihala Beach County Park**. The park can also be approached from the Migrant Trap and Hammond Marina. A bicycle path connects the Migrant Trap and Whiting Park. Another lakefront park, **Jeorse Park** (V5) in East Chicago, can be found off Cline Avenue near the casino.

Trails—Walk the paths in Whiting Park, along the lakefront from Whihala Beach toward the Hammond Marina and Migrant Trap and check both sides of the cement wall between Whiting Park and the railroad tracks.

Telephone—(219)853-6436 for Hammond City Hall and (219)659-7700 for Whiting City Hall.

Wolf Lake (V2)

W S S F

This shallow lake straddles the Illinois-Indiana State Line. The water at the north end of the lake, visible from the Indiana shores, may stay open in winter. Waterfowl, shorebirds, gulls, and terns visit Wolf Lake in season.

Directions—Drive into **Forsythe Park** by turning west from Calumet Avenue (Rte. 41) on 119th Street. To bird the southern portion of the lake, take 129th Street west of Calumet and Sheffield Avenue. Access from the Illinois side is discussed earlier in this guide.

Lake George (V3)

W S S F

A shallow lake, woods, and **Calumet College Marsh** attract a large variety of migrants. Waterfowl, shorebirds, gulls, and terns utilize the open water and muddy borders. King Rails have nested here. The woods draw so many passerines that local birders refer to the area as a migrant trap.

Directions—Turn east off of Calumet Avenue on 125th Street. The lake is visible on both sides of the causeway. Continue east to the marshy areas and woods. Park along the road.

Trails—Walk around the west side of the NIPSCO substation to reach the north woods or walk south of 125th Street to the south woods. The marsh can be approached from 129th Street.

Note—The woods are best birded with companions.

Hammond Cinder Flats (V4)

W S S F

When water is present, this site can be very productive. Herons, Black-bellied Plovers, Pectoral Sandpipers, and dowitchers are regular visitors in the spring, late summer, and fall. More unusual shorebirds have included American Avocet, Hudsonian Godwit, and Western Sandpiper.

Directions—Turn east from Calumet Avenue (Rte. 41) on 141st Street in Hammond. Continue toward Columbia Street and park on the north side. The area is fenced, but it can be birded from 141st Street with a telescope.

Roxana Pond (V6)

This wetland was originally a bend in the Grand Calumet River. As one manifestation of how profoundly development has altered the natural character of this area, the widely varying water levels of Roxana Pond are now more affected by discharges from local companies than they are by precipitation. When water levels are low (not often in recent years), large numbers of shorebirds visit the pond and the river banks. American Avocet, Marbled Godwit, Stilt Sandpiper, Long-billed Dowitcher, and Red-necked Phalarope have all been recorded. Black-crowned Night-Herons and egrets are seen in summer and waterfowl in spring and fall.

Directions—From Indianapolis Boulevard, turn west on Roxana Street in Hammond. Park on Roxana Street to bird the pond and continue west on Roxana under the I-90 overpass to bird the Grand Calumet River. Turn onto Roosevelt west to Molseberger where you can park and walk west to view the river from the park.

Trails—View the pond from Roxana Street. Good views of the river and its mudflats can also be had from the Columbia Avenue bridge. The river to the west is open during winter due to the release of warm water from the wastewater treatment plant on the north side of the river west of Columbia Avenue and I-90.

Gibson Woods Nature Preserve (V7)

Before northwestern Lake County was covered with steel mills, residential areas, and other development, 150 narrow beach ridges and swales crossed the lacustrine plain from Gary to the Illinois border. Botanists reveled in its richness; nearly 30 species of orchids were among the many native plants that grew here in profusion. One early writer attested to the many times he slaked his thirst with the pure water that filled the shallow swales, an act that today would be masochistic if not suicidal. About 1,000 acres (a tenth of the original area) remain intact, and much of this is in public ownership. One of the largest tracts is Gibson Woods Nature Preserve with 131 acres of wooded sand dunes, swales, sand savanna, sand prairie and shrubby swamps. A large variety of birds have been seen here due to the many habitats clustered into a small area. In winter, feeders attract Tufted Titmice, Red-winged Blackbirds, Common Grackles, and American Tree Sparrows. Early spring brings hawks, Sandhill Cranes, American Woodcock, and sparrows. Later, in May, waves of warblers filter through the open woods, and Olive-sided Flycatchers sally from the branches of dead trees.

Directions—Exit onto Kennedy or Cline Avenue from I-80/94, go north to 169th Street, then go east (west if coming from Cline) to the light at Parrish Avenue. Turn north to the Nature Center at the end of the road.

Trails—The preserve has a self-guided nature trail and a wheelchair-accessible boardwalk. Four trails are marked and loop through the preserve. There are two half-mile trails, a one-mile trail and a two-mile trail. A naturalist leads scheduled hikes into the preserve.

Note—Gates are open from 9:00 A.M. to 5:00 P.M. daily (4:00 in the winter). The Nature Center is open from 11:00 A.M. to 5:00 P.M. (4:00 in the winter). Closed on Christmas, New Year's Day, and Easter holidays.

Owned By—Lake County Parks and Recreation Department

Telephone—(219)844-3188

Miller Beach (V8)

W S S F

A bird flying as far south as it can over Lake Michigan or its shoreline will eventually hit Miller Beach. This "funnel effect" has made Miller Beach one of the best locations in the region for observing lake birds. From August to November, the passage of each cold front brings the promise for an exciting day of birding punctuated with unexpected finds. Among the most coveted of local birds are any of the three jaegers. Though rare, more jaegers have been seen here than at any other place in the Midwest. Look for them, along with Little Gulls, Sabine's Gulls, and Black-legged Kittiwakes, on days with northwest winds from September through November. Loons (including Red-throated), grebes, ducks, American Avocets, Whimbrels, terns, and some truly amazing rarities (Northern Gannet and Gull-billed Tern) have all been seen here in the fall.

Directions—Turn north from Rte. 12 onto Lake Street in Miller. Pass Miller Woods and the Indiana Dunes National Lakeshore's Paul H. Douglas Center for Environmental Education. Drive to the large lot at the end of Lake Street or, for a more sheltered and elevated position from which to view the lake, park one mile east in **Marquette Park** near the concession stand. To reach the concession stand, drive east from Lake Street on Hemlock to Grand Street and turn north. Grand Street winds through Marquette Park before coming to Montgomery. Turn left on Montgomery and follow it to the concession stand on the right. Park along the road.

Trails—Bird from your car in the lot or walk the beach west of Lake Street toward the steel mills. A fee is charged to park in the Lake Street parking lot from Memorial Day to Labor Day.

Owned By—National Park Service, U.S. Department of the Interior

Oak Ridge Prairie County Park (V11)

W S S F

Within this 700-acre recreational park, birders can sample a wide range of habitats: wet prairie, marsh, oak savanna, sedge meadow, canals, and ponds. In early spring, this is a good spot to hear Common Snipe and American Woodcock in courtship flights. American Bitterns have been seen here; Wood Ducks are seen frequently; and Black-crowned Night-Herons visit the area in summer. Nesting species include Cardina Wren, Sedge Wren, Scarlet Tanager, Eastern Towhee, Grasshopper Sparrow, Henslow's Sparrow, Eastern Meadowlark, and Orchard Oriole. The Sedge Wrens have been found in several locations: near the pond along the entrance road, south of the fishing pond, in the wet prairie, and in the sedge meadows.

Directions—Turn east on Main Street from U.S. 41 toward Griffith and continue to Colfax Street. Turn south on Colfax for .5 mile to the entrance on the east side of the road. The park is open from 7:00 A.M. to 5:00 P.M., and a parking fee is charged in summer.

Owned By—Lake County Park and Recreation Department

Telephone—(219)844-3188

Note—**Hoosier Prairie State Nature Preserve** (V10) is two miles west of the park on the south side of Main Street and Kennedy Avenue. There is a small parking lot at the entrance for the hiking trail.

Deep River County Park (V12)

These 1,000 acres of forest along the Deep River attract passerines during migration and entice a good variety of species to nest, including Barred Owl, Blue-gray Gnatcatcher, Eastern Bluebird, Wood Thrush, and Prothonotary Warbler. An early spring treat are the Sandhill Cranes that often pass over on their way to breeding grounds farther north.

Directions—Turn north from Highway 30 on Randolph Street (approximately one mile east of Highway 51). At the intersection of 73rd (Old Lincoln Highway), turn east for approximately one mile to the park at the intersection of County Line Road.

Trails—A trail follows the west bank of the river from the visitors' center to the north entrance of the park on Ainsworth. Canoes can be launched near the visitors' center and taken out at the north end of the park.

Owned By—Lake County Park and Recreation Department

Telephone—(219)947-1958

Cedar Lake (T1)

A glacial lake nestled in the Valparaiso Moraine, Cedar Lake has long been known as a favorite resting area for waterfowl. Loons, grebes, and ducks continue to flock here, particularly in spring.

Directions—Exit U.S. Highway 41 at the exit for Cedar Lake on 133rd Avenue and travel east until you see the lake on the south side of the road. Park in the pull-off across the street from the funeral home or further east in the lot for the Chamber of Commerce.

Trails—Bird from the parking areas. In the early morning, waterfowl can be quite close to the shore.

Grand Kankakee Marsh County Park (T7)

Attitudes toward wetlands have changed over the years more than they have toward probably any other landscape type. In the beginning of this century, tremendous effort was devoted to draining the Kankakee Marshes. Now, almost 100 years later, a growing list of agencies are converting former marshland back to its original character. Lake County owns 1900 acres of wetlands here, including a restored area known as Goose Lake. Waterfowl and shorebirds use this county park during migration, as do numerous passerines. (Hunting is allowed here, so spring birding is best.) Species recorded in the nesting season have included Ruby-throated Hummingbird, Pileated Woodpecker, Wood Thrush, Yellow-throated Warbler, and American Redstart.

Directions—Exit I-65 east at Rte. 2 to Clay Street (Range Line Road), then south for five miles. Park in the lot on the left before crossing the river. The park is open from 7 A.M. to dusk from January through September.

Trails—Walk along the dike road to the east or west of the lot and look for levees heading north from the dike. A scope will be helpful during waterfowl and shore-bird season. The Lake County naturalist from Gibson Woods leads trips here.

Owned By—Lake County Park and Recreation Department

Telephone—(219)552-9614

Schneider Sod Farms (T2)

W S S F

These sod farms may be the most productive for shorebirds in the entire region. American Golden-Plovers, Upland Sandpipers, and Buff-breasted Sandpipers appear in good numbers from August through early October.

Directions—The sod farms are on both sides of Rte. 41 north of the town of Schneider and the Kankakee River.

Trails—No trails exist. Observe from Rte. 41 which has wide shoulders at this point.

Porter County

Port of Indiana (W3)

W S S F

Amidst this large commercial complex, several opportunities exist for the birder during fall and winter. The harbor provides sanctuary for wintering diving ducks (including a King Eider now and then), the breakwaters hold potential for Purple Sandpipers and Snowy Owls, and the location is a good vantage point to observe migrants. Lapland Longspurs and Snow Buntings are often seen, as they glean spilled cereal from the access road. But the best birding is during fall migration, when strong winds propel flocks of birds across the viewing area. Western Grebe, all three scoters, Parasitic Jaeger, and Black-legged Kittiwake have all been seen here.

Directions—Turn north from I-94 on Rte. 249 in Burns Harbor. Stop at the state police guardhouse at the port entrance, proceed due north to the public access area right of the grain elevators, and park. Special permission may be required to enter.

Trails—Check the access road, the harbor, and the breakwaters. The point between two channels is where most of the rarities have been seen, but it may be difficult to reach.

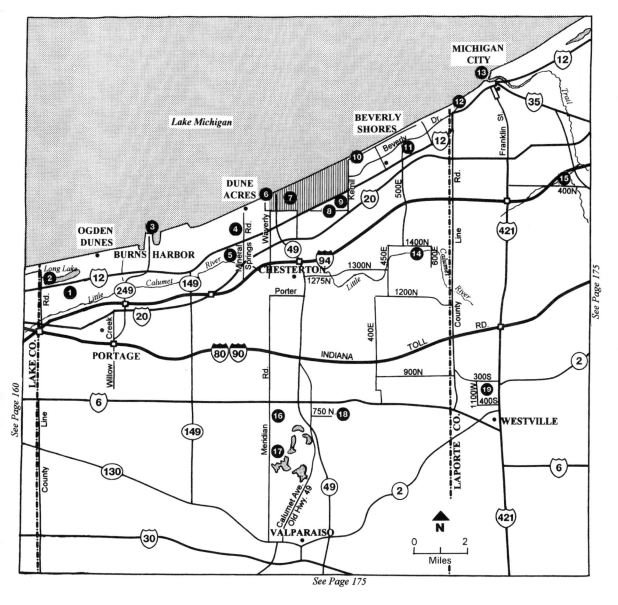

Map W **North Central Indiana**

1. Inland Marsh *p. 169*
2. Long Lake *p. 168*
3. Port of Indiana *p. 166*
4. Cowles Bog *p. 169*
5. Little Calumet River Trail *p. 169*
6. Johnson Beach *p. 170*
7. Indiana Dunes State Park *p. 170*
8. Ly-co-ki-we Horse and Hiking Trail *p. 171*
9. Indiana Dunes National Lakeshore Visitor Center *p. 168*
10. Kemil Beach *p. 172*
11. Beverly Shores *p. 172*
12. Mount Baldy *p. 172*
13. Michigan City Harbor *p. 173*
14. Heron Rookery *p. 172*
15. Creek Ridge County Park *p. 174*
16. Sunset Hill Farm County Park *p. 173*
17. Valparaiso Lakes *p. 173*
18. Moraine State Nature Preserve *p. 173*
19. Bluhm Woods County Park *p. 174*

Indiana Dunes National Lakeshore

W S S F

With the exception of Lake Michigan, the Indiana Dunes are the most significant natural features existing in the Chicago region. They dominate the cultural, economic, and physical landscape of this part of Indiana. For a century, this small province has captured the human imagination. The body of work produced by the scientists, writers, poets, musicians, painters, and sculptors celebrating the dunes is exceeded locally only by that focusing on Chicago itself. Such feelings spurred the battle to save the dunes, a struggle that has been called "perhaps the most savage conservation-industry confrontation in history" (Platt in Engel 1983, 5). And when Congress did finally preserve what was left of the dunes, it required an unprecedented act: the creation of the country's first urban national park. Until the recent establishment of Midewin National Grassland, the Indian Dunes National Lakeshore was the only large federally owned park in the entire region. From a biological perspective, this attention is clearly warranted. Except for the much larger Grand Canyon and Great Smokey Mountains, this 12,000-acre National Lakeshore harbors more species of plants than any other unit within the national park system.

Virtually every plant community known from the Chicago region is represented in the dunes: open beach, dune, lake, marsh, tamarack bog, forested fen, prairie, savanna, and several types of forest. This great diversity can be explained in part by the area's nearness to Lake Michigan (moister and milder than inland locations) and the variations in topography, soil, fire frequency, and age of land forms.

The **Dorothy Buell Memorial Visitor Center** (W9) for the National Lakeshore is a good place to become oriented to the area, to pick up a map, and to purchase reference books. The Visitor Center is located on Highway 12 at Kemil Road, 3 miles east of Highway 49. Naturalists from the National Lakeshore offer scheduled hikes; call to learn the schedule of excursions. It is also worth noting that the dunes can be accessed via the South Shore Railroad, which has several stops within the National Lakeshore.

Owned By—All sites are owned by the National Park Service except for the Indiana Dunes State Park, which is owned by the Indiana Department of Natural Resources, and most of Beverly Shores, which is privately held.

Telephone—(219)926-7561

Long Lake (W2)   This interdunal lake is the largest inland water body within the National Lakeshore. Waterfowl alight on the open water during migration, and Great Egrets, Least Bitterns, and both night-herons patrol the shoreline or emergent vegetation of the shallows. The dunes, pines, and shrubs to the north of the lake have drawn numerous passerines in migration, and Northern Goshawk, Northern Shrike, Pine Grosbeaks, Red Crossbills, White-winged Crossbills, and Common Redpolls during the winter.

Directions—Drive north from I-80/94 or the toll road on Rte. 51 to Rte. 20. Go east one mile to County Line Road and turn north two miles to the entrance to **West Beach** and Long Lake. After passing the entrance kiosk (where a fee is charged in summer), park in the first lot on the right for Long Lake. For the beach and dunes, continue to the main lot. The beach is not visible from the main parking lot.

Trails—Walk back along the entrance road from the Long Lake parking lot for .3 mile to reach the Long Lake Trail which ascends the dune at the west end of the lake and affords a good view of the lake. The light is best in the afternoon from this point. For the beach, park in the main lot and walk 200 yards north.

Inland Marsh (W1)   Migrants find shelter and forage within this area of vegetated dunes and small wetlands. American Bitterns have been seen in the marsh in May, and Eastern Bluebirds generally nest here. A good assortment of butterflies are present in mid-summer.

Directions—Exit I-80/94 or the toll road at Rte. 51. Go north on Rte. 51 to its end at Highway 20 and turn right. After one mile, turn left (north) at County Line Road, then right (east) at Rte. 12. After two miles, look for the parking lot for the marsh on the right.

Trails—A half-mile trail goes south from the lot through a flat, grassy area before entering the dunes. Two trails form loops through the dunes and swales to the west. The yellow trail is one mile long; the red trail extends one mile further. The trails lead to isolated areas, so it may be prudent to bird with a companion.

Cowles Bog (W4)   When is a bog not really a bog? When it is Cowles Bog, which is actually a forested fen watered by a calcium-rich spring rather than acidic surface water. Although there are tamaracks present, there is virtually no sphagnum moss, a key bog indicator. This botanically important site hosts numerous migrants and nesters. Two potentially excellent places to observe migrant passerines are the trail through the woods and the brushy areas near the fence and dike. Virginia Rail, American Woodcock, White-eyed Vireo, Sedge Wren, Marsh Wren, Yellow-breasted Chat, and Swamp Sparrow can be seen and heard in the breeding season from Mineral Springs Road.

Directions—Drive north of Highway 12 on Mineral Springs Road toward Dune Acres. Park at the north end of the road near the guard station for Dune Acres or in the lot at the railroad tracks, but be aware that some cars left in the lot near the railroad tracks have been vandalized.

Trails—Walk west from either parking spot. Immediately west of the guard station, a marker commemorates Professor H.C. Cowles, for whom the bog is named. A trail proceeds west through woods, past the bog, and ultimately ends up at Lake Michigan. From the railroad lot, the trail skirts the NIPSCO property and fence. The NIPSCO settling ponds are beyond the dike, and recent brush removal makes it possible to see the ponds from the dike. The trail then circles north to meet the trail between the beach and the guard station. From the trail's end at the beach, it is possible to walk east three miles to Indiana Dunes State Park. A walk along Mineral Springs Road is a good way to view the marsh.

Note—Poison sumac grows quite close to Mineral Springs Road.

Little Calumet River Trail (W5)   The Little Calumet River winds through woods at the base of the Lake Border Upland. The combination of water, trees, and shelter attracts many land bird migrants.

Directions—Exit I-80/94 at Rte. 20 and drive east 1.2 miles. Turn north on Howe Road and park at Baily Homestead. Parking is also available at Chellberg Farm on Mineral Springs Road north of Rte. 20.

Trails—The Little Calumet River Trail heads west along the river, turns north, and circles back along a moraine.

Johnson Beach (W6)   One of the lakeshore's most popular birding activities is the spring hawk watch, and the dune at Johnson Beach is a favorite viewing area. March 1 through May 15 is the period of peak raptor migration; nine of the ten largest flights ever recorded here occurred in April. Hawk watchers have identi-fied a total of 16 species, of which Sharp-shinned, Broad-winged, and Red-tailed make up over 70 percent of the individuals tallied. The flights tend to concen-trate in a narrow band within a mile of the lakefront, so if you are out on a sunny spring day with southerly winds and aren't seeing any hawks, consider moving to another location. Two good options are the parking lots on Highway 12 at Highway 49 or Tremont Road farther east.

Directions—Turn north from Rte. 12 on Waverly Road, the first street west of Rte. 49. The dune at the north end of the road is your goal. To park, jog left, then right on Wabash around the base of the dune. For the closest parking, turn right on Duneland and park at the end. Walk up to the top of the dune.

Indiana Dunes State Park (W7)   There were two major battles in the effort to save the dunes. The most recent battle culminated in the establishment of the na-tional lakeshore in 1966. The first began in 1914 and ended in 1927 when the State of Indiana created the Indiana Dunes State Park, which now contains 2,182 acres. Because it has been under protection for so long, the park encompasses "the greatest expanse of undamaged dunes remaining in the state" (Brock 1997, 15). No other portion of the dunes can rival the state park for the number and variety of migrant and nesting passerines. In fact, of the 39 species of warblers recorded in the dunes, only Kirtland's has never been seen within the state park. Trails #2 and #10 traverse the wet woods and marshy border that are the park's most productive birding habitats. Trail #10 can be particularly rewarding on cool spring mornings when warblers are forced to glean insects from low-lying vegeta-tion. An early summer's hike along these trails will likely produce many of the following nesting species: Red-shouldered Hawk, Broad-winged Hawk, Barred Owl, Red-headed Woodpecker, Acadian Flycatcher, White-eyed Vireo, Veery, Blue-winged Warbler, Northern Parula, Cerulean Warbler, Prothonotary Warbler (along Trail #10 where the footbridge crosses the marsh), Louisiana Waterthrush, Hooded Warbler, and Canada Warbler. A third trail, Trail #9, follows the sandy dune tops and is quite strenuous to hike, but it passes through the blowouts in-habited by the region's only colony of nesting Prairie Warblers.

Directions—To reach Indiana Dunes State Park, go north of Chesterton on High-way 49, cross Highways 20 and 12, and go directly into the State Park. An en-trance fee is charged. For the woods and marsh, park in the lot near the Wilson Shelter east of the gatehouse. The eastern two-thirds of the park is a dedicated nature preserve. For the beach, park at the end of the road past the gatehouse.

Trails—Pick up a State Park trail map from the gatehouse. The picnic area at the Wilson Shelter can be good for thrushes, sparrows, and even a Solitary Sand-

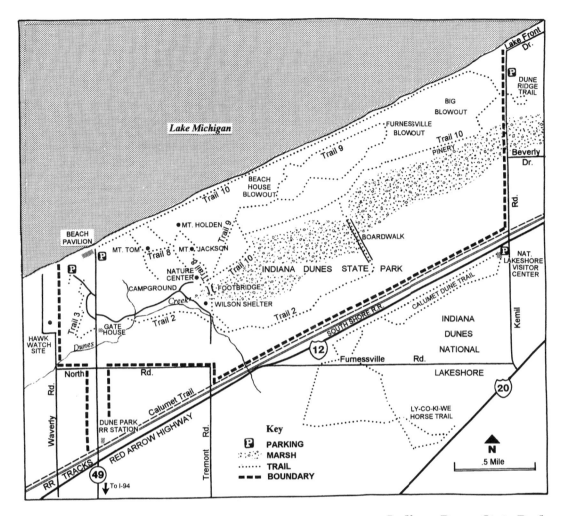

Indiana Dunes State Park

piper in spring migration. Trail #2 is south of the shelter and passes through tall trees and wooded swampland. Most birders walk east on Trail #2, but a walk to the west can be good, too. After walking east for a little over a mile, the trail turns north, crosses the interdunal marsh, and meets Trail #10. Turn left to loop back through marsh, wooded swamp, and savanna habitats on your way to your car. For a longer and more rigorous return, turn right on Trail #10 and circle back on Trail #9. This route takes you up steep, wooded dunes, follows dune crests to blowouts, and has spectacular views of treetops and the lake. The sandy blowouts can be approached from the beach too. Park in the lot near the beach house and walk along the shore to the east until you see the blowouts.

Owned By—State of Indiana Department of Natural Resources.

Telephone—(219)926-1952

Note—Rubber boots may be necessary in spring on Trail #2. Just south of the park, the **Ly-co-ki-we Horse and Hiking Trail** (W8) provides access to an area of woods where Hooded Warblers have nested. Turn north from Rte. 20 just east of Kemil Road.

Beverly Shores (W11)   This small residential community represents a private enclave surrounded by the National Lakeshore, but the adjacent woods, marsh, dunes, and beaches are also referred to as Beverly Shores. The roads that cross the dunes south of Lake Front Drive are good places to see the spring migrants that are often abundant in the adjacent woods. Species that nest or summer in the shrubby marsh along Beverly Drive include Alder Flycatcher, Willow Flycatcher, White-eyed Vireo, Common Yellowthroat, Yellow-breasted Chat, and Swamp Sparrow. Woodland birds such as Acadian Flycatchers and Cerulean Warblers have nested near Kemil Road, while Prairie Warblers and Field Sparrows nest just west of the parking lot in the State Park and are sometimes seen from Kemil Road. Look for Red-throated Loons offshore in October.

Directions—For **Kemil Beach** (W10) and Lake View picnic area, park north of Highway 12 in the parking lots for the National Lakeshore on Kemil Road. For the woods near Highway 12, park in the lot on Kemil Avenue on the south side of Rte. 12 and cross the highway to bird. Birding along Lake Front Drive and Beverly Drive can be rewarding, but parking is prohibited and strictly enforced.

Trails—For lakefront birding, walk to the beach from the lots on Kemil Road or Central Road. For the interdunal marsh, drive or walk along Beverly Drive; the area east of Broadway is particularly good for Yellow-breasted Chats, and an artesian well near the west end of Beverly Drive is attractive to birds during dry periods. South of the intersection of Beverly Drive and Kemil Road, old road-ways provide access to a dense woodland. The Calumet Bike Trail provides access to this area from the south. A spring keeps the wooded area quite wet. The north/south roadway between Beverly Drive and the railroad tracks just east of Kemil Road is referred to as Owl Road because owl walks are scheduled there by the naturalists at the National Lakeshore.

Mount Baldy (W12)   Mt. Baldy, on the eastern edge of the National Lakeshore and barely within LaPorte County, is another high dune that affords views of spring raptor flights. Impressive movements of Sandhill Cranes, Northern Flickers, blue jays, crows, and swallows can also be observed from the summit. In recent summers, a Chuck-Will's-Widow has taken up residence near the parking lot, where it can be heard calling at dawn and dusk.

Directions—Get off I-94 at Highway 421. Head north through Michigan City, then west on Highway 12. Look for the sign for Mt. Baldy on the north side of the street.

Trails—Follow the stairway and walking trail to the dune crest.

Indiana Dunes Heron Rookery (W14)   The wet woods along the Little Calumet River have had a colony of Great Blue Herons for 60 years. While the herons have colonized the eastern portion of the tract, the best area for birding is actually on the west end. Spring migrants crowd trees along the river, and Yellow-throated Warblers and American Redstarts nest here in greater numbers than anywhere else in the dunes. Starring three species of trillium, one of the National Lakeshore's most dazzling exhibitions of spring wildflowers blankets the floor of its beech-maple forest.

Directions—Drive south from Rte. 20 on 500E (Lake Shore County Road) to 1400N. Turn right on 1400N and go .5 mile west to 450E. Turn south for .75 mile and park in the small lot across from the intersection of 450E and 1325N.

Trails—Walk east on the 1.5-mile trail on the south side of the Little Calumet River from 450E to 600E. The trail is quite close to the stream and walking can be rough.

Note—A parking lot is located on the east side of the preserve on 600E, but access from this lot may disturb the herons.

Moraine State Nature Preserve (W18)

The 474-acres within this preserve represent an excellent example of a morainal landscape, including wooded ridges, a spatterdock-covered pond, and marshy potholes. Winter birding often turns up Eastern Screech-Owls, Great Horned Owls, and Barred Owls, and summer visits have yielded Blue-winged, Golden-winged, and Chestnut-sided Warblers. During spring and fall the woods come alive with migrant land birds, most notably an excellent variety of warblers that has included Worm-eating (regularly found in the ravine north of the road), Connecticut, and Mourning (attracted to the edges of the small ponds).

Directions—Drive south of Rte. 6 on Old 49 (North Calumet) to 750N (Meska). Faith Lutheran Church is on the corner. Turn east and continue straight ahead, past the dangerous curve to the right. No parking facilities or trails have been developed. Call the Department of Natural Resources ahead of time to request permission to enter.

Trails—Bird from the road or park in the few small pullouts along the road. Beyond the first small pond on the right, a trail to the north circles around a swamp and returns to the road beyond the second pond on the right.

Owned By—State of Indiana Department of Natural Resources

Telephone—(219)843-5012

Note—**Sunset Hill Park** (W16) on Meridian Road south of Highway 6 has grassland species, and **Valparaiso Lakes** (W17) further south on Meridian Road have waterfowl in migration.

LaPorte County

Michigan City (W13)

Migrant passerines of all kinds gather in the scanty vegetation of the harbor area which includes Washington Park and the Northern Indiana Public Service Company (NIPSCO) facility. (The NIPSCO property is particularly good on spring days with westerly winds.) Flocks of water birds stream by the lighthouse, while others linger in the harbor or on the breakwaters and beaches. Birders are advised to scrutinize all of this bird activity closely, for no other site in the state boasts more rarities. The list of unexpected birds has included Magnificent Frigatebird, Slaty-backed Gull, Royal Tern, Long-billed Murrelet, and Kirtland's Warbler. The parade of rarities has stretched over decades and has led Ken Brock to proclaim Michigan City Harbor the best birding location in the entire state. No season holds a monopoly on rarities, but autumn seems to yield more than other seasons.

Directions—Take I-94 to U.S. 421 and go north to Michigan City and the lake. Highway 421 becomes Franklin and ends at a "T." Turn right at 4th Street, then left on Pine Street which leads to Washington Park. Keep right and park in the lot between the yacht basin and beach (a parking fee is charged during summer months). For the NIPSCO site, park west of the Coast Guard and Department of Natural Resources offices and east of the NIPSCO entrance. To view Trail Creek further upstream, go east on U.S. 12 and U.S. 35. Turn left (north) on 8th Street and view the creek from the bridge on "E" Street.

Trails—If the weather and waves permit, walk out to the lighthouse at the end of the jetty for views of birds flying past or perched on the outer breakwater. Scan the Washington Park beach east of the harbor entrance and the NIPSCO beach west of the harbor entrance. Be sure to check the inlet at Trail Creek near the Coast Guard Station for unusual waterfowl. For the NIPSCO site, walk down to the path between the NIPSCO plant and Trail Creek and walk around the fence toward the beach.

Note—Several areas near Michigan City offer birding opportunities including **Creek Ridge County Park** (W15) on County Road 400N (Kiefer Road) east of Rte. 421, **Bluhm Woods County Park** (W19) on County Road 1100W between 300S and 400S west of Rte. 421, and **Hudson Lake** (T13) north of Rte. 20 off of Emery Road at Chicago.

La Porte Lakes (T11)

W S S F

Pine Lake, Stone Lake and Horseshoe Lake are all glacial lakes of the Valparaiso Moraine. During migration they attract such species as Common Loons, Horned Grebes, Canvasbacks, Redheads, Ring-necked Ducks, and Buffleheads. Black Terns have nested near Horseshoe Lake.

Directions—From LaPorte, drive north on Rte. 35; turn west on Waverly Road to view Pine Lake and Stone Lake. Soldiers Memorial Park has two parking areas off of Waverly Road and another access off of Rte. 35 south of Waverly. Further north, Severs Road west of Rte. 35 provides an overlook for Pine Lake. Clear Lake is east of Rte. 39 on McClung Road; take Truesdale Avenue east from Pine lake Avenue to reach Fox Park. Horseshoe Lake is north of LaPorte and east of Rte. 39 on County Road 300N.

Trails—Soldiers Memorial Park offers public access for Pine Lake and Stone Lake. Fox Park provides access to Clear Lake.

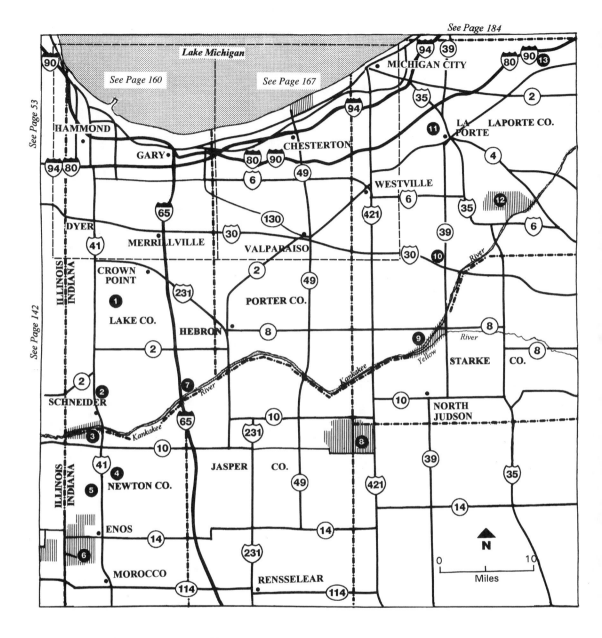

Map T **Greater Indiana**

1. Cedar Lake *p. 165*

2. Schneider Sod Farms *p. 166*

3. LaSalle Fish and Game Area *p. 177*

4. Beaver Lake Prairie Chicken Refuge *p. 178*

5. Newton County Fields *p. 178*

6. Willow Slough Fish and Wildlife Area *p. 178*

7. Grand Kankakee Marsh County Park *p. 165*

8. Jasper-Pulaski Fish and Wildlife Area *p. 180*

9. Kankakee Fish and Wildlife Area *p. 182*

10. Shamrock Turf Nurseries *p. 177*

11. La Porte Lakes *p. 174*

12. Kingsbury Fish and Wildlife Area *p. 176*

13. Hudson Lake *p. 174*

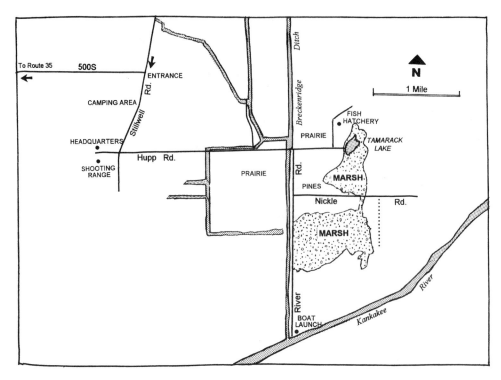

Kingsbury Fish and Wildlife Area

Kingsbury Fish and Wildlife Area (T12)

W S S F

This is one of five Indiana Fish and Wildlife areas located in what used to be the great marsh of the Kankakee River, an area estimated to have been between 500,000 and 1,000,000 acres in size. The 5,000 acres that are protected here include marsh, grasslands, and evergreen plantings. In years when winter finches are in the region, Red Crossbills and Common Redpolls take advantage of the conifers. In the same season, Northern Harriers, Rough-legged Hawks, and Northern Shrikes patrol the fields for rodents. They are joined by Short-eared Owls which like to squat atop the "shredded wheat" hay stacks. With the onset of spring comes migrating waterfowl, herons, shorebirds (including Wilson's Phalaropes in late April–May), and sparrows. In March, White-fronted Geese are a good possibility. Nesting species include Chestnut-sided Warbler, Cerulean Warbler, Black-and-white Warbler, American Redstart, and Prothonotary Warbler in the river bottoms and Sedge Wren (best seen in June), Grasshopper Sparrow, and Bobolink in the grasslands. It is one of the best sites in the region to find both species of cuckoos during the breeding season.

Directions—The area is 10 miles southeast of the town of LaPorte. Drive east from Highway 35 on 500S. After about 4.5 miles, the road ends at a "T" intersection. Turn right and cross the railroad tracks to enter the area. The check station is west on Hupp Road; good birding areas can be found by driving two miles east on Hupp Road and then south on River Road to the big marsh and the Kankakee River. Hunting is allowed in season, so a call to the Fish and Wildlife Division is advisable to find what areas are affected.

Trails—This area was used as a U.S. Army ammunition depot during World War II. Contaminated areas are marked with signs and are off limits. Drive the roads slowly and bird from the road.

Owned By—State of Indiana Department of Natural Resources

Telephone—(219)393-3612

Shamrock Turf Nurseries (T10)

From August through September, as they head to wintering grounds on the South American pampas, American Golden-Plovers, Upland Sandpipers, Buff-breasted Sandpipers, and other shorebirds stop here to feed in the damp short grass of the sod farms. Around the same time, large flocks of Brewer's Blackbirds can often be found, making these sod farms Indiana's best location for this species. Western Meadowlarks nest 1 mile east of the sod farms.

Directions—Turn north from Highway 30 onto Rte. 39 and park immediately in the lot on the west side for the turf farm, or pull off onto gravel County Road 1300S on the north side of the sod farm to view the sod farms and creek.

Owned By—privately owned

Newton County

LaSalle Fish and Game Area (T3)

On the Indiana/Illinois state line, this preserve is at the west end of the stretch of the Kankakee River that runs through a ditch. (In the old days before it was straightened, the river didn't run as much as it lollygagged.) The 3,648 acres of bottomland woods, old meanders, and overgrown fields support a host of interesting species. The wet woods have nesting Red-shouldered Hawks, Barred Owls, Pileated Woodpeckers (sporadic and difficult to find), and lots of Prothonotary Warblers. Drier areas near the campground attract Red-headed Woodpeckers and Eastern Bluebirds, and the areas near headquarters may yield nesting Bell's Vireos, Yellow-breasted Chats, and Orchard Orioles.

Directions—Go 1.5 miles west of Highway 41 on Highway 10 to 415W. Turn north to the park headquarters. Another entrance, particularly good for Barred Owls and Prothonotary Warblers, is the first road that goes west from Rte. 41 south of the Kankakee River. A site map is available. Hunting is allowed here, and access is restricted from October through January.

Trails—Trails exist near headquarters, and walking is possible in the campground and fishing area along the river.

Owned By—State of Indiana Department of Natural Resources

Telephone—(219)992-3019

Beaver Lake Prairie Chicken Refuge (T4)

W S S F

Although the last of Indiana's Greater Prairie Chickens has been gone for over 25 years, a wide variety of grassland birds still inhabits this 640-acre nature preserve. It is particularly good during May and June when Northern Bobwhite, Northern Mockingbirds, Henslow's Sparrows, Blue Grosbeaks, Dickcissels, and Orchard Orioles are easily found.

Directions—From Rte. 41, go east on 400N for one mile and then north on 200W. The preserve is on the east side of the road.

Trails—Most birders cover the area from road 200W which runs along the preserve's western edge. Contact the manager at the LaSalle State Fish and Game Area for permission to enter the preserve.

Newton County Fields (T5)

W S S F

South of the Kankakee River, on either side of Rte. 41, there are fallow fields and flooded agricultural land that provide excellent birding. In the winter, hawks are easily found along the roadsides, and Horned Larks, Lapland Longspurs, and Snow Buntings congregate on gravel roads. By early April, waterfowl, Yellow Rails (very rarely seen), American Golden-Plovers, and Smith's Longspurs are passing through. During the nesting season, Northern Bobwhites, Upland Sandpipers, Bell's Vireos, Yellow-breasted Chats, Dickcissels, and such sparrows as Field, Savannah, Grasshopper, and Henslow's are usually in the vicinity.

Directions—Look for little-used roads to the east and west of Highway 41 south of the Kankakee River. Drive 7.5 miles south of the river to County Road 400N. Turn west to County Road 400W which parallels Highway 41 or turn east to County Road 250W and 200W which parallel Highway 41. The most productive fields are south of County Road 400W. An area of sand hills is north and east of this area. By driving north to County Road 800N and turning two–three miles east, you enter an area of sand hills and oaks that have had nesting Northern Mockingbirds, Summer Tanagers, Lark Sparrows, and Orchard Orioles.

Willow Slough Fish and Wildlife Area (T6)

W S S F

One of the most celebrated portions of the old Kankakee Marsh was Newton County's Beaver Lake which once had 36,000 acres of open water and marsh. Descriptions exist of the lake being covered with waterfowl, of how a flight of teal four or five feet deep took up acres of sky. It was the last place in the Midwest where Trumpeter Swans nested. When it was finally drained in 1917, an event that took place during the nesting season, the stench of rotting fish and young waterfowl filled the air for miles around. Willow Slough Fish and Wildlife Area occupies 9,600 acres of this former wonderland.

Managed for fishing and hunting, the lakes, ditches, marshes, fields, upland woods, and conifers also provide excellent birding. Raptors, including an infrequent Golden Eagle, work the fields in winter. In spring and fall, Sandhill Cranes can be heard overhead while rails and Sharp-tailed Sparrows skulk in

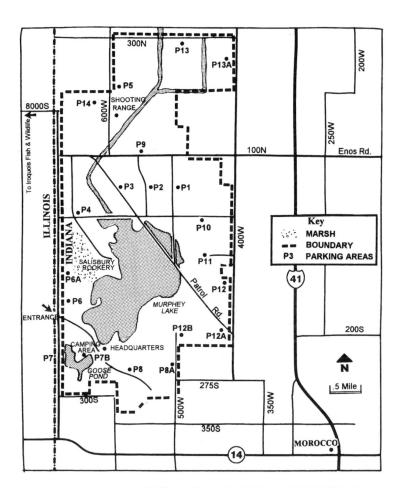

Willow Slough Fish and Wildlife Area

grassy wetlands. While the Fish and Wildlife area is but a pale imitation of old Beaver Lake, a taste of the bygone days might be imagined when huge numbers of waterfowl, mostly Ring-necked Ducks, scaup, and other diving ducks, stop at Murphy's Lake during migration. Spring is generally best for birding as there is no hunting at that time of year. Adding to the variety, Horned Grebes, Tundra Swans, Ospreys, and Bald Eagles are also occasionally encountered. Nesting species are another major draw for birders. Specialties include Whip-poor-wills, Summer Tanagers, Lark Sparrows, and Blue Grosbeaks. Good locations to look for those species are the fields and woods near parking lots #5 and #14. Nesting Hooded Mergansers, Bell's Vireos, Yellow-breasted Chats, and Orchard Orioles may be found in the marsh and hedgerows along Patrol Road and south on 600W. Nesting Lark, Grasshopper, and Henslow's Sparrows can be found in fields near parking lot #9. Herons and dabbling ducks collect in the Salisbury Rookery, and Northern Bobwhite live amidst the fencerows throughout the area.

Directions—Turn off Highway 41 onto Rte. 14 in Enos westbound. This road becomes County Road 100N. The intersection of County Road 400W marks the eastern boundary of Willow Slough. The wildlife area extends three miles west, two miles north and four miles south of this point. To obtain a map, continue three miles west to State Line Road and turn left. Look for the paved road on the left after 2.5 miles. The headquarters for the wildlife area are in the boathouse near the fishing and camping area on Murphy's Lake. Parking Lot #5 is one mile north of County Road 100N on the east side of County Road 600W, and Parking Lot #14 is one mile further north on the west side of 600W. Patrol Road cuts diagonally across the area from County Road 100N on the northwest to County Road 400W. Salisbury Rookery is close to Parking Lot #6A on State Line Road .75 mile north of the road to the headquarters. The signs for the parking lots may be missing. Hunting and fishing are allowed here, so wildlife becomes wary in season.

Trails—The north-south dike near the refuge headquarters is good in spring and fall for sparrows.

Owned By—State of Indiana Department of Natural Resources

Telephone—(219)285-2704

Note—Directly across the state line in Illinois is the **Hooper Branch Savanna Nature Preserve** and the **Iroquois County Conservation Area** (S9), sites with diverse habitats that offer excellent birding. To reach the area, drive north of County 100N on State Line Road. Turn west at the first opportunity, then south in two miles. (Further description under Kankakee County.)

Jasper County

Jasper-Pulaski Fish and Wildlife Area (T8)

W S S F

In Aldo Leopold's "Marshland Elegy," one of his most beautiful essays, he writes about Sandhill Cranes: "When we hear his call we hear no mere bird. He is the symbol of our untamable past, of that incredible sweep of millennia which underlies and conditions the daily affairs of birds and men" (1970). No where else east of Nebraska do more of these magnificent creatures congregate than at Jasper-Pulaski, a refuge occupying a part of the former Beaver Lake basin. Peak numbers of cranes occur during the fall migration from mid-October through November and usually number around 15,000 individuals. The highest single-day count was established in November 1991, when observers counted 32,000 birds. Smaller, although still impressive, concentrations form in March and April. In spring a visit to the observation deck toward early dusk will reveal hundreds of cranes dancing and leap-frogging over each other in graceful courting ceremonies. (One local ornithologist of the 1850s thought their movements "uncouth.") Most of the cranes leave the refuge to feed in neighboring fields during the day. As the sun wanes, more and more cranes return to Jasper-Pulaski. Ducks dart across the field and alight in small ponds, as Northern Harriers and Rough-legged Hawks search for prey emboldened by the approaching darkness. Deer emerge from the woods to graze with the cranes. And when the light is gone and the birder heads back to her vehicle, she is serenaded by the music of cranes, woodcock, Eastern Chorus Frogs, and Spring Peepers. Do not miss this spectacle.

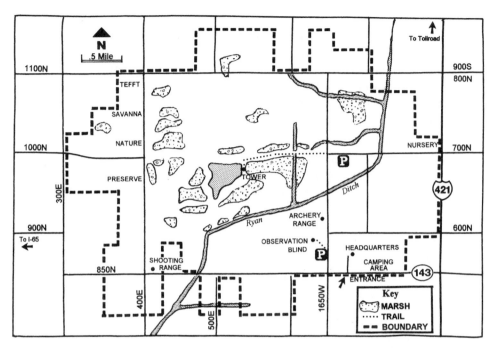

Jasper-Pulaski Fish and Wildlife Area

Of course, Jasper-Pulaski is not just cranes and waterfowl. Both species of eagles are seen here with some frequency, probably lured by the abundance of waterfowl that use the refuge. The woods and waters ensure the presence of many migrant passerines, including good numbers of Rusty Blackbirds that can be found along the dike. An interesting selection of "southern" specialties have summered here, including Northern Bobwhite, Barred Owl, Whip-poor-will, Loggerhead Shrike, Bell's Vireo, Tufted Titmouse, and Carolina Wren. Summer Tanagers have nested in the Tefft Savanna Nature Preserve, and Pileated Woodpeckers have been seen in the same area.

Directions—Turn west from Highway 421 onto Rte. 143 (County Road 850N) and look for the entrance on your right. If approaching from the west, go east on Indiana Highway 10 from I-65 and continue straight ahead for 14 miles until that road ends. (The highway designation changes along the way to Highway 110 and County Road 900N.) At the "T" intersection with County Road 300E, turn right and go about .25 mile, then turn left near the small power station on County Road 850N (the road is not marked). Continue driving east to the entrance to the park about 3.5 miles ahead on the left. Visitors are requested to check in at the check station where maps are available and the latest count of cranes is posted. The parking area for the crane observation tower is just north of County Road 850N on 1650W. The parking area for the waterfowl observation tower is one mile further north. Drive north on County Road 1650W across Ryan ditch to County Road 700N. Turn west (left) on what becomes a dike and park in the lot at the end. (Cars are not permitted to drive west on 700N during hunting season weekends; use the lot just east of the intersection and walk west to the tower.) For the Tefft Savanna Nature Preserve, park in lot #6 on County

Road 400E or in lot #11 on County Road 1100N. County Road 1100N continues to the east where it is marked 900S or 800N. Grasslands can be seen near parking lot #5 on this road. The shooting range is located on County Road 850N west of the crane observation tower. Bell's Vireos have nested in this area. Hunting is allowed throughout the area; if you want to visit when no hunting is allowed, call the refuge to see when the various seasons begin and end.

Trails—A short (.25 mile) walk leads to the crane observation tower overlooking the refuge, but cranes can usually be seen from the parking area. A telescope is helpful. The waterfowl observation tower is 1.25 miles west of the intersection of 700N and 1650W. The road to the tower is on an elevated dike which passes bottomland forest, a wooded swamp, and lakes on the way to the observation tower. This road is closed to vehicles on weekends during hunting season and must be walked at those times. Most wetland areas are off-limits to the general public. The Tefft Nature Preserve in the northwest section of the park has no trails, but fire lanes can be walked.

Owned By—State of Indiana Department of Natural Resources

Telephone—(219)843-4841

Starke County

Kankakee Fish and Wildlife Area (T9)

W S S F

In the 1870s, the English Lake area of the Kankakee Marsh was nationally famous for its hunting and fishing. It featured the lake and the point where the Kankakee River is joined by the Yellow River, its second largest tributary. Several lodges catered to wealthy sportsmen from around the country. The Kankakee Fish and Wildlife Area now protects 4,095 acres of this stretch. During early spring when there are thousands of ducks and geese in the air and water, the birder might, with a little effort, believe that he has been transported back in time to when the great marsh was still a forbidding place teeming with life. Good birding is available at other times of the year as well. Barred Owls and Pileated Woodpeckers are permanent residents in the woods, and an excellent assortment of woodland passerines nest here including Brown Creeper, Veery, Wood Thrush, Yellow-throated Warbler, Cerulean Warbler, and Prothonotary Warbler.

Directions—From Rte. 30, go south on Highway 39. The wildlife area extends west and south from the intersection of Highways 39 and 8. The headquarters building will be found if you continue 4 miles south to Toto Road (County Road 300S), turn right, and proceed 2.3 miles west. The fields around the building and to the west are very good in late winter to early spring for water birds, and the pond behing the building attracts waterfowl. Other fields may be seen from County Road 400W north of Toto Road (a dead end), and in LaPorte County north of the Kankakee River. To reach the fields north of the river, drive 3 miles west from the intersection of Highways 39 and 8 to County Road 500W. Turn left and go south for 1 mile to County Road 2100S. Turn right and go west for 1 mile to 600W, then south for 1 mile to 2200S. This intersection is known to attract large numbers of waterfowl. For the forest, turn west from the intersection of Highways 39 and 8 and park in the lot on the north side of the road. A trail makes a loop north of the lot in habitat that is excellent for woodland birds. The wooded river banks south of Highway 8 can be birded from the 10-mile drive that makes a loop between Highways 39 and 8. Call headquarters first if the road conditions are questionable. By turning south onto the loop from Highway 8, you drive the loop counter-clockwise. By turning west onto the loop from Highway 39, you drive it clockwise. A shortcut allows you to drive only 5 miles if desired, but the southern end of the 10-mile loop is a good location for Pileated Woodpeckers.

Note—The gravel pit on the east side of Highway 39 just south of Rte. 30 also attracts waterfowl in migration.

Owned By—State of Indiana Department of Natural Resources

Telephone—(219)896-3522

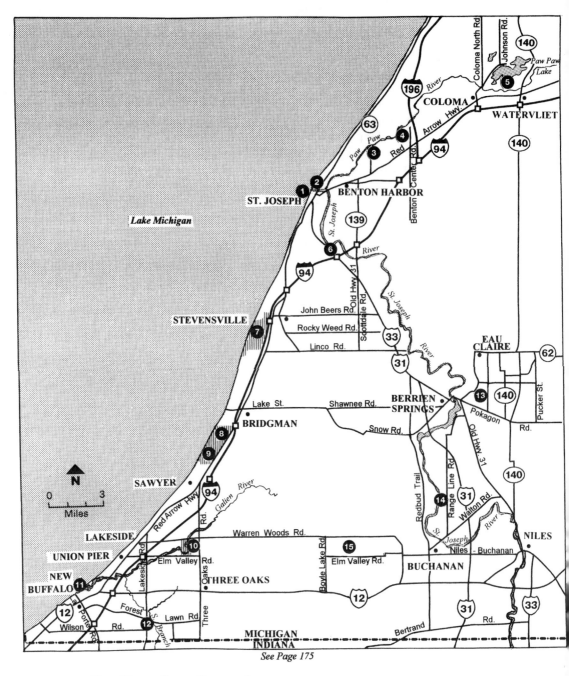

See Page 175

Map X **Berrien County**

1. St. Joseph/Benton Harbor *p. 185*
2. Jean Klock Park *p. 185*
3. Brown Sanctuary *p. 186*
4. Sarett Nature Center *p. 186*
5. Paw Paw Lake *p. 186*
6. Riverview Park *p. 185*
7. Grand Mere State Park *p. 187*
8. Floral Road Entrance *p. 188*
9. Warren Dunes State Park *p. 188*
10. Warren Woods State Park *p. 189*
11. New Buffalo Harbor *p. 190*
12. Forest Lawn and Lakeside Roads *p. 190*
13. Love Creek County Park and Nature Center *p. 191*
14. Fernwood Botanic Garden and Nature Center *p. 191*
15. Mud Lake Bog Nature Preserve *p. 191*

Berrien County, Michigan

St. Joseph/Benton Harbor beaches (X1)

W S S F

The beaches and piers at the mouth of the St. Joseph River on Lake Michigan provide excellent birding. Loons, grebes, geese, ducks, raptors, terns, and an assortment of passerines pass these vantage points in great numbers. Shorebirds often stop to rest on the beaches during summer and early fall. Check out any gull concentrations, as rarities seem partial to this small area. The list of unusual gulls recorded here include Black-headed, California, and Sabine's. Northern Mockingbirds have been seen regularly in the Shoreham residential area just south of St. Joseph east of Lakeshore Drive.

Directions—Go to St. Joseph by exiting I-94 at Exit 23 (St. Joseph Business I-94 Exit) or Exit 27 (Niles/M63 Exit). The two roads meet and become Main Street and continue to the lakefront. Bird the beach and piers on both sides of the St. Joseph River. On the south side of the river, Silver Beach County Park can be found by following the signs on Main Street. For the north side of the river continue north on Main Street (M63). As soon as you cross the river, turn right on Upton Drive. Make a hard left turn onto Prospect Street after crossing the railroad tracks and follow the signs to **Tiscornia Beach Park**.

Trails—Scan the lake and nearby shoreline for birds and walk out onto at least one of the piers for a closer look. Be sure to scan the beaches north and south of the parks.

Note—Jean Klock Park (X2), one mile north of the mouth of the St. Joseph River, has a beach, a pond, and a wetland on Lake Michigan. Besides the migrant water birds, the park has produced some remarkable flights of passerines: 3,000 Bank Swallows (July), 5,000 Cedar Waxwings (May), 5,000 American Goldfinches (October), and 2,000 Evening Grosbeaks (November). Rails and other marsh birds may be found in the wetland area and pond across from the main park entrance. To get there from Tiscornia Beach Park, go to the main gate for the park on Jean Drive north of Upton. The wetland area is directly east of the main gate. If the gate is closed, try parking in the lot for the Benton Harbor Water Department at the end of North Upton Drive and walking in through the gate at the west end of the lot (the north end of Ridgeway Drive). Walk the beach and search the marshes back from the beach.

Riverview Park (X6) on the St. Joseph River can have good concentrations of migrants. Turn off from Rte. 33 into the park.

Sarett Nature Center (X4)

Nestled in the bends of the Paw Paw River, Sarett Nature Center has a wide diversity of habitats with 362 acres of woods, old fields, alder thickets, wetlands, marshes, and bogs. Wintering birds have included Virginia Rail, Common Snipe, Winter Wren, White-crowned Sparrow, and Common Redpoll. American Woodcock perform courtship displays in the old field areas in March and April. During spring and fall migrations a variety of sparrows can be found in the old fields, and warblers are often active along the wooded ridge. Birds that have bred here include Virginia Rails and Soras in the wetlands, Prothonotary Warblers along the Paw Paw River at the end of the River Trail, and White-eyed Vireos and Yellow-breasted Chats in the shrubby marsh. Take advantage of the boardwalk that allows easy exploration of the various wetland communities. A trip in early fall should yield blooming Fringed Gentians and Grass of Parnassus, two stunners that are increasingly difficult to find.

Directions—Exit I-94 at Exit 34 (the I-196/Grand Rapids Exit) and follow I-196 north one mile to Exit 1 (Red Arrow Highway). Turn left (west) onto Red Arrow Highway and proceed .5 mile to Benton Center Road. Turn right and go north one mile to the entrance to the Nature Center, which has feeders and exhibits. Trails are always open, but the Nature Center is closed on Monday, open from 9:00 A.M. to 5:00 P.M. Tuesday through Friday, 10:00–5:00 on Saturday, and 1:00–5:00 on Sunday. Picnics and pets are prohibited on the site.

Trails—Trail maps are available at the Nature Center. Over five miles of trails loop through different habitats and include a boardwalk and observation platforms.

Note—The **Brown Sanctuary** (X3), downstream on the Paw Paw River, is managed by Sarett Nature Center. Turn north from the Red Arrow Highway at Crystal Avenue (at the east side of the Benton Harbor Airport) and look for the entrance sign in .75 mile. A list of birds seen is kept in a box near the entrance.

Owned By—Michigan Audubon Society

Telephone—(616)927-4832

Paw Paw Lake (X5)

Berrien County's largest inland lake offers a place of rest for migrating loons, swans, diving ducks, and gulls. Once the temperature warms, however, motorized boat activity precludes worthwhile birding.

Directions—Exit I-94 at Coloma and drive north through town to Paw Paw Street and veer left on Paw Paw Lake Road. Before reaching Johnson Road, watch for a view of Paw Paw Lake on your right. By turning right on Shoreview, you can view the lake from the parking lot of the Paw Paw Yacht Club. Further east, views can be had from the parking lot for Lakeside Landing Restaurant or from the Paw Paw Lake Marina. To see **Little Paw Paw Lake**, drive north on Johnson Road for .6 mile to Helen Street and drive west to the small park at the end of the road.

Owned By—Private property surrounds most of Paw Paw Lake, but there is a small Department of Natural Resources public access at the east end of the lake.

Grand Mere State Park (X7)

The three lakes of Grand Mere State Park were formed out of a bay in glacial Lake Chicago. South Lake is the shallowest and has the most emergent vegetation, and North Lake is the largest and least vegetated. In between them in location, size, and succession is Middle Lake. The lakes attract a variety of ducks in April. A strip of dunes separates the lakes from Lake Michigan, except where a channel enables North Lake to discharge into Lake Michigan. The vegetation within this unique complex of wetlands and dunes is extraordinarily diverse and includes one of the region's few hemlock swamps, a plant community that reaches its southern limit in the Midwest in Berrien County. Given the northern plant species that thrive here, it is not surprising that several northern birds nest as well. These include such warblers as Black-throated Green, Black-and-white (near Baldtop Dune), and Canada (along the entrance road). But birds with southern affinities have also summered here including White-eyed Vireo, Worm-eating Warbler (in the wooded dunes), and Hooded Warbler. Birding Grand Mere during the breeding seasons can be exciting, but the mix of habitats next to the lake also makes this an excellent spot for migrants, from waterfowl to raptors to passerines. Birders with the stamina to hike the dune crests will be rewarded by good views of tree top warblers in migration.

Directions—Exit I-94 at Exit 22 for Stevensville. Turn west toward the dunes. To scan North and Middle Lake, go straight ahead. North Lake's parking area will be on the right; Middle Lake is further west and can be viewed from the boat launch ramp on the left. To visit the dunes and wooded areas, turn south on Thornton Road which parallels I-94. After about .5 mile, turn right at the park entrance and follow the gravel road .5 mile west to the main parking area, South Lake, and trails. For the southern portion, continue south on Thornton Road to Willow, turn right, and continue to the end of Wishart. The state park is open to hunters in the fall; deer firearm season is generally November 15–30.

Trails—The main parking area and the gravel road leading to it are often good for woodland birds in migration. To see Lake Michigan, open dunes, and two of the inland lakes, follow the paved trail leading west from the parking area. A short paved path leads to South Lake on the left. After scanning the lake, return to the paved trail and continue west to the second bench on the right near the end of the pavement. Follow the unpaved trail on the right that leads up a hill to Baldtop Dune. Keep to the right and follow the trail back east along the edge of Middle Lake. This trail has a few short, steep climbs in sand and ultimately leads back to the parking area. From Wishart Road on the southern end, the trail to the north leads to the main parking lot. Many unmarked trails exist, so it is easy to get lost.

Owned By—Administered by Warren Dunes State Park

Telephone—(616)426-4013

Warren Dunes State Park (X9)

W S S F

E. K. Warren made his fortune by developing and selling turkey-bone corsets late in the nineteenth century. Among his philanthropic activities was purchasing the lakefront property that became Warren Dunes State Park. Over the years the property has expanded to 1,950 acres. Warren believed that the site would benefit large numbers of people by providing a place for education and recreation. Warren Dunes now has parking spaces for 2,500 cars, and is very popular among summer beach-goers and campers. The wooded and open dunes, wetlands, and beaches are also very popular among avian visitors, including waterfowl, hawks, shorebirds, gulls, terns, and woodland passerines.

The **Floral Road** (X8) area with the Yellow Birch Trail and Goldenrod Trail is at the undeveloped northern end of Warren Dunes State Park and is a favorite of birders. Spring and fall migration can be outstanding for a host of species. What makes this part of the park most interesting, however, is the list of nesters, including Whip-poor-wills, Pileated Woodpeckers, Blue-winged Warblers, Chestnut-sided Warblers, Black-throated Green Warblers, Black-and-white Warblers, Ovenbirds, and Hooded Warblers.

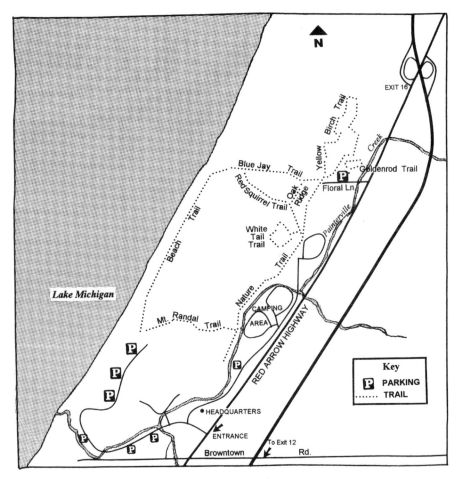

Warren Dunes State Park

Directions—For the main entrance to the park, exit I-94 at Exit 12. Go west on Sawyer Road, then north on Red Arrow Highway to the park. An entry fee is required. The entrance road leads to the beach. Park at the beach, in the lots along the main road or in the lot for the nature trail. To access the undeveloped Floral Road section, drive 1.5 miles north of the entrance road to Floral Road and turn west. Alternatively, get off I-94 at Exit 16 and drive south on the Red Arrow Highway for .7 mile. Turn west on Floral Road and park in the small lot on the right.

Trails—The beach area is easily accessed from the beach parking lots. A variety of trails and roads lead through the park and campground. The undeveloped northern end of the park, Yellow Birch Trail, and Goldenrod Trail can be reached by walking on the nature trail north of the campground. From the Floral Road parking lot, walk to the end of the road and follow Yellow Birch Trail to the north along the base of the wooded dunes, around a wetland and swamp area, and back to the entry road, or take the Goldenrod Trail through an old field area. A boardwalk has been constructed over 500 feet of the wettest area of the Yellow Birch Trail. Two other trails begin in the parking area. One trail heads west over the dunes to the beach, and another trail goes south to the state park campground and meets other trails in Warren Dunes State Park.

Note—A list of recent bird sightings is maintained at the trailheads for the Yellow Birch and Goldenrod Trails. Boots may be necessary in spring to walk on the Yellow Birch Trail. Hunting is allowed in the northern part of the park in the fall; the deer firearm season is generally November 15–30. Call headquarters to learn if hunting is taking place before visiting.

Owned By—Michigan Department of Natural Resources

Telephone—(616)426-4013

Warren Woods State Park (X10)

Besides Warren Dunes, E. K. Warren purchased this piece of property that was also destined to be a state park. It was a 150-acre wooded tract along the shores of the Galien River. Within this larger area, expanded now to 480 acres, there is a 39-acre parcel on the north side of the river that may be the finest example of a virgin beech-maple forest existing anywhere in the world. Scientists have been studying the forest for over 60 years, for living laboratories of this quality are difficult to find. The big timber makes this the best place in the region to find Pileated Woodpeckers, although they are elusive. Barred Owls also live here year round. Large numbers of warblers and other passerines forage in the vegetation during migration, and a good selection stay to breed. Among them are Acadian Flycatchers, Yellow-throated Vireos, Tufted Titmice, Carolina Wrens, Cerulean Warblers, Ovenbirds, Louisiana Waterthrushes, and Hooded Warblers.

W S S F

Directions—Exit I-94 at Exit 6 (Union Pier Road Exit). Turn east on what becomes Elm Valley Road and continue for approximately 2.6 miles. An inconspicuous gravel road on the left is the official entrance to the park. A .25-mile drive leads to a small parking lot with a pit toilet and picnic table. A daily vehicle entry permit or annual state park pass is required to park in the lot. A better option is to go to the north end of the park. To do so, continue .5 mile east to Three Oaks Road, turn north (left), and proceed 1 mile to Warren Woods Road. At the intersection, turn west (left) and park along the road after crossing the bridge.

Trails—Trails wind through the forest and along the river. A footbridge connecting the northern part of the park to the southern part where the parking lot exists was destroyed in 1995. Call to see if it has been replaced or walk south from Warren Woods Road on the west side of the river to bird the old-growth woods located on the north side of the river.

Owned By—Administered by Warren Dunes State Park

Telephone—(616)426-4013

New Buffalo Harbor (X11)

W S S F

This harbor complex at the mouth of the Galien River is a fine location for birders wishing to see waterfowl and shorebirds migrating along the Lake Michigan shoreline. The small harbor, jetties, and beaches north of the harbor and on both sides of the inner harbor provide habitat for resting grebes (including Red-necked on occasion), ducks, shorebirds, gulls, and terns. Over ten species of gulls including Laughing Gull and Little Gull are recorded here each year. In November and December, look for Harlequin Ducks, scoters, Purple Sandpipers (on the jetties), and Snowy Owls.

Directions—Exit I-94 at Exit 1 (New Buffalo/La Porte Road Exit) and turn west. Proceed straight ahead on Whittaker, cross the Galien River, and park in the public beach parking lot. A fee is charged during summer months.

Trails—Observation decks are located on the sand dune. The cattail marsh is located north of the marina, but access is a problem. You can canoe into the marsh by putting in at the boat launch near the beach.

Forest Lawn and Lakeside Roads (X12)

W S S F

An excellent variety of spring migrants and nesting woodland species can be seen in these woods on the floodplain of the south branch of the Galien River. Red-shouldered Hawks have nested near Forest Lawn Road, and the tall Sycamore trees in this area are the best place in Michigan to see Yellow-throated Warblers. Other summering species to look for include Acadian Flycatcher, Blue-gray Gnatcatcher, Northern Parula, Cerulean Warbler, and Louisiana Waterthrush.

Directions—Exit I-94 at Exit 1. Go south, then immediately east on Wilson Road. At the end of Wilson, turn north on Lakeside Road. The intersection with Forest Lawn Road is less than one mile north. Turn east on Forest Lawn and park near the bridge where it crosses the river. After birding this area, return to Lakeside Road, go north, and park near the bridge. Another nearby area to search is Witt Road 1.5 miles east of this intersection.

Trails—Walk along the road to bird as the surrounding property is private.

Love Creek County Park and Nature Center (X13)

The park surrounding the Nature Center contains beech-maple forest, old fields, and a creek. Migrant passerines pass through in substantial numbers. Acadian Flycatchers and Louisiana Waterthrushes have nested in the riparian woods, Blue-winged Warblers are found in the old fields, and Pileated Woodpeckers are seen year-round in the timbered portions. Though rare, Ruffed Grouse have been seen here. Call the Nature Center for information on the latest sightings and other local birding information. The Berrien Birding Club meets here.

W S S F

Directions—From Berrien Springs, cross the St. Joseph River and turn left at the sign for Dean's Hill Road. Follow the signs for the Park three miles east on Huckleberry Road.

Note—A trail map can be picked up at the Nature Center, which is closed on Monday. A $5.00 fee is charged for non-resident parking.

Trails—Twelve trails with a total of five miles have been established.

Owned By—Berrien County Parks and Recreation Department

Telephone—(616)471-2617

Fernwood Botanic Garden and Nature Center (X14)

Land-bird migrants are abundant in this 105 acres of woods, gardens, pond, grassland, and arboretum on the east bank of the St. Joseph River. Hummingbirds frequent the gardens, and numerous other birds gather at the feeders. A regional specialty that has been found along the Wilderness Trail is Ruffed Grouse. Although once present in heavy woods throughout the Chicago Region, this species is now locally extirpated except for Berrien County.

W S S F

Directions—Drive northeast from Buchanan on Walton Road, then north on Range Line road to the garden at 13988 Range Line Road. The facility charges a $3.00 admission fee. It is open 9:00 A.M. to 5:00 P.M. Monday through Saturday and 12 noon to 5:00 Sunday. The Nature Center is open at noon on weekdays. A tea room serves lunch from 11:30 A.M. to 2:00 P.M. every day except Monday.

Trails—Visit the Nature Center where a naturalist can provide information about local birds. A walk from the Nature Center down to the river and along Wilderness Path circles through the wooded portion of the property.

Owned By—Fernwood Botanic Garden

Telephone—(616)695-6491

Note—Mud Lake Bog (X15), a tamarack bog on Elm Valley Road west of Buchanan, is worth a visit during summer for its unusual plants and nesting Red-shouldered Hawks, Barred Owls, and marsh birds. Sandhill Cranes have nested here since 1997.

Birds of the Chicago Region

The following **species** are generally recorded every year or two in the Chicago area. Certain species such as Pine Grosbeak and White-winged Crossbill occur less frequently but are included because when present they are often in sizable numbers. The **time periods** given indicate when the birds are most abundant. This does not mean they do not occur at other times. For example, most puddle ducks are described as migrants but many regularly winter and some have summered. Indeed, out-of-season records exist for most species.

There have been many attempts to define terms of **abundance** through some means of quantification (see Brock 1997, 26–29). That has not been done here. Rather, the terms are comparative and derive meaning from their placement on a continuum defined at either extreme:

> Abundant *(large numbers in appropriate habitat during designated period[s])*
> Common
> Fairly Common
> Uncommon
> Rare
> Very Rare *(expected once every year or two)*

Finally, for species of special interest that can be found in specific locations, representative sites of frequent but not guaranteed occurrence are given. Species are listed in taxonomic order as published in the 41st supplement (July 1997) of the American Ornithologists Union Checklist of North American Birds.

Red-throated Loon—Rare in April and November in the open water and harbors of Lake Michigan. Most regular off Beverly Shores in early November.

Common Loon—Common to fairly common from late March through late April and late October to late November in the open water and harbors of Lake Michigan and other lakes.

Pied-billed Grebe—Fairly common migrant from late March to mid-May and mid-September to mid-October on any size body of water; less common on Lake Michigan. Uncommon and local nester where there is a mixture of open water and emergent vegetation.

Horned Grebe—Fairly common from mid-March to late April and early October through early December. Occurs on Lake Michigan as well as on inland lakes.

Red-necked Grebe—Rare in March and April and mid-October through early December. Occurs almost exclusively in open water and harbors of Lake Michigan.

Eared Grebe—Rare from mid-March through late May and from late September through late December. Occurs in open water and harbors of Lake Michigan, as well as in small lakes.

Western Grebe—Rare from late October through early December. Occurs in open water and harbors of Lake Michigan.

American White Pelican—Rare, but increasing in frequency. Records exist for all seasons, although almost never in winter.

Double-crested Cormorant—Common to fairly common in all seasons except winter. Population has increased markedly in recent years. Nests at Willow Slough, Lake Renwick, Saganashkee Slough, and Baker's Lake. Largest colony is at Riverdale Quarry.

American Bittern—Uncommon to rare migrant from early April through mid-May and early September through late October. Increasingly rare nester in large marshes. Migrants often seen on Chicago lakefront. Suggested breeding locations include Illinois Beach State Park and Goose Lake Prairie.

Least Bittern—Uncommon to rare migrant and nester occurring between late April and early September. Most often found in marshes where they are known to breed. These include Cuba Marsh, Des Plaines Wetlands Demonstration Project, Powderhorn Marsh, McKee Marsh, Illinois Beach State Park, Nelson Lake, Glacial Park, Long Lake, and Willow Slough.

Great Blue Heron—Common to fairly common in all seasons except winter, although increasingly birds are staying throughout the year. Found in virtually all areas near water. Nests in colonies of varying size; among the best known are those at Baker's Lake, Lake Renwick, Heron Rookery, and Riverdale Quarry.

Great Egret—Fairly common to uncommon in all seasons except winter. Distributed widely in habitats with shallow water: rivers, lakes, flooded fields, sloughs, and marshes. Nests in colonies, including those at Lake Renwick, Baker's Lake, Willow Slough, and Indian Ridge Marsh.

Snowy Egret—Rare from late April through mid-September. Not known to nest. Fraternizes with other herons and most reliably found at Lake Renwick and wetlands in the Calumet region.

Little Blue Heron—Rare from late April through early September, with frequency peaking in late summer. Fraternizes with other herons and regularly seen at Lake Renwick and wetlands in the Palos and Calumet areas.

Tricolored Heron—Very Rare from May through September. Most frequently seen in Lake Calumet area and along lakefront.

Cattle Egret—Rare from April through October. During migration appears on lakefront, in flooded fields, and in sloughs. Breeding colony at Lake Renwick.

Green Heron—Fairly common from late April through late September. Does not nest in colonies, thus is relatively inconspicuous but widespread on edges of marshes, lakes, and waterways.

Black-crowned Night-Heron—Fairly common from late April through late September. Nests in colonies, but forages widely. Large nesting colonies at Lake Renwick and Big Marsh.

Yellow-crowned Night-Heron—Rare from late April through September. Nests in wooded wetlands in small colonies or individually. Most often seen as post-breeder in Calumet area, Lockport Lock and Dam, and other open wetlands.

Turkey Vulture—Fairly common to uncommon from early March through late October. Most common in southern part of region, although increasing everywhere. During migration can be seen anywhere; during breeding season prefers less developed areas.

Greater White-fronted Goose—Rare from ice break-up through mid-April and then again from October through November. Largest numbers are in spring; also increasing during winter. Occurs with other geese. Good places to look include Fermilab, Kemper Lakes, Lake Renwick, Willow Slough, Crabtree Lake, and Kankakee Fish and Wildlife.

Snow Goose—Uncommon from mid-March through mid-April and then from early October through end of November. Individuals also seen amidst winter flocks of Canada Geese. Good places to look include Fermilab, refuges in Kankakee River basin, Palos, and along Lake Michigan.

Ross's Goose—Rare from mid-March through mid-April and then from mid-October through early winter. Often found alone and seems to be increasing. Places to look include Fermilab, refuges in Kankakee Basin, and Chicago Botanic Gardens.

Canada Goose—Abundant year-round resident.

Brant—Very rare from mid-March through mid-April and then from mid-October through the end of December. Occurs mostly in harbors and open waters of Lake Michigan.

Mute Swan—Introduced nuisance increasing in numbers and locations. Can be seen throughout the year.

Tundra Swan—Uncommon to rare in March and early April and then November through late December. Most often seen in fall migrating overhead or on medium- or large-sized lakes. Lake Michigan shoreline is particularly good.

Wood Duck—Common from mid-March through mid-November in many lakes, ponds, sloughs, streams, and rivers with wooded edges.

Gadwall—Fairly common from mid-March to late April and then common from late September through mid-November. Prefers shallow lakes, sloughs, and flooded fields.

Eurasian Wigeon—Very rare from mid-March through late April and then from early October through late November. Usually found with American Wigeon.

American Wigeon—Common from early March to late April and then from mid-September through late November. Prefers sloughs, shallow lakes, and flooded fields.

American Black Duck—Common from early March thorough mid-April and then from early November through early December. Found on rivers, marshes, and lakes, including Lake Michigan. A few birds are usually found in flocks of wintering mallards.

Mallard—Abundant year-round resident wherever there is water.

Blue-winged Teal—Common to fairly common from mid-April through mid-October. Prefers ponds, sloughs, and flooded fields.

Northern Shoveler—Common from early March through late April and then mid-September through early November. Prefers shallow lakes, sloughs, and flooded fields.

Northern Pintail—Uncommon from early March through early April and then from late October through mid-November. Prefers shallow lakes and flooded fields.

Green-winged Teal—Fairly common from early March through late April and then from beginning of October through end of November. Most often found in shallow lakes, flooded fields, and sloughs.

Canvasback—Uncommon from when ice leaves until mid-April and then from late October through late November. Prefers deep lakes, including Lake Michigan.

Redhead—Fairly common from early March through late April and then from early October through early December. Prefers flooded fields, sloughs, and lakes (shallow, deep, and Lake Michigan).

Ring-necked Duck—Common from early March through late April and then early October through late November. Prefers flooded fields, sloughs, and lakes (shallow, deep, and less frequently, Lake Michigan).

Greater Scaup—Common from late February through early April and then from mid-October through early December. Often large flocks remain through winter. Found mostly on open water and harbors of Lake Michigan.

Lesser Scaup—Common from late February through late April and then from mid-October through mid-December. Found on wide range of waters—flooded fields, sloughs, and lakes (shallow, deep, and Lake Michigan).

King Eider—Very rare from mid-October through early March in harbors and open waters of Lake Michigan.

Harlequin Duck—Rare from mid-October through early April in harbors and open waters of Lake Michigan.

Surf Scoter—Uncommon to rare from early October through late March in harbors and open waters of Lake Michigan.

White-winged Scoter—Fairly common to uncommon from late February through mid-March and uncommon from early October through late December. Restricted to harbors and open waters of Lake Michigan.

Black Scoter—Uncommon to rare from early October through late March in harbors and open waters of Lake Michigan.

Oldsquaw—On western shore of Lake Michigan (south to southern Chicago), uncommon to rare from late October through early April. On eastern shore of Lake Michigan, rare from early November through end of March. Numbers fluctuate from year to year but has decreased markedly in the Chicago region.

Bufflehead—Common from late February through mid-April and from mid-October through early December. Prefers lakes, sloughs, and Lake Michigan, but occasionally occurs on rivers.

Common Goldeneye—Abundant to common from early November through mid-March. Prefers Lake Michigan, but also found on rivers and other lakes.

Hooded Merganser—Fairly common to uncommon from mid-March through late April and from early October through late November. Prefers lakes (shallow and deep) and sloughs. In fall and winter, also seen on Lake Michigan and rivers.

Common Merganser—Common to fairly common from early November through mid-April on Lake Michigan, large lakes, and large rivers.

Red-breasted Merganser—Abundant to common from mid-March through early May and from mid-October through early December on Lake Michigan, large lakes, and large rivers.

Ruddy Duck—Common to fairly common from mid-March through early May and from late September through late November on lakes (shallow and deep), sloughs, and less frequently, Lake Michigan.

Osprey—Uncommon to rare from early April through mid-May and uncommon from early September through late October. Most often seen migrating along Lake Michigan shoreline, but also seen in vicinity of rivers and other lakes.

Bald Eagle—Very uncommon to rare from mid-October through mid-April. Most often seen along Lake Michigan shoreline, but also seen in vicinity of other large bodies of water. A virtual certainty at Heidecke Lake in winter.

Northern Harrier—Uncommon from early March through late April and fairly common from early September through early November. Prefers fields, marshes, and Lake Michigan shoreline.

Sharp-shinned Hawk—Fairly common to uncommon in April and common to fairly common from early September through early November. One of the three most abundant hawks at raptor lookouts.

Cooper's Hawk—Fairly common to uncommon from mid-April through late November. Has increased dramatically in recent years and can be found in all but the most heavily urbanized areas.

Goshawk—Except in invasion years, rare from mid-October through late March. Most often seen migrating along lakefront and in woods with evergreens.

Red-shouldered Hawk—Uncommon migrant in the Indiana Dunes during March and early April. Rare permanent resident elsewhere in riparian woods.

Broad-winged Hawk—Fairly common from mid-April through mid-May and from early September through early October. Rare summer resident in areas of extensive woods.

Swainson's Hawk—Small numbers nest in Kane County and southwestern McHenry County. Generally present from early April through late September, but most easily seen from May through July. Known to occur in vicinity of Coral Woods, Hampshire Forest Preserve, and Johnson's Mound Forest Preserve.

Red-tailed Hawk—Common permanent resident throughout area, except most urbanized portions.

Rough-legged Hawk—Fairly common to uncommon from late November through late March in areas of extensive open land and as a migrant along Lake Michigan shoreline. Abundance varies from year to year.

Golden Eagle—Very rare from early March through mid-April and rare from mid-October through mid-December. Most often seen migrating along Lake Michigan shoreline and at refuges in the Kankakee River area.

American Kestrel—Fairly common to uncommon permanent resident in open areas throughout region, including heavily urbanized portions.

Merlin—Rare from late March through mid-April and uncommon from mid-September through mid-October. Most often seen migrating along Lake Michigan shoreline in fall.

Peregrine Falcon—Uncommon from mid-September through late October. With recent introductions and establishment of a breeding population, this falcon has become a rare permanent resident. Most often seen in downtown Chicago or migrating along Lake Michigan shoreline.

Gray Partridge—Introduced gamebird that is rare and decreasing in numbers as urbanization continues moving west. Restricted to open areas in western and northwestern portions of region.

Ring-necked Pheasant—Introduced gamebird losing ground to development but still widely found in open areas throughout region.

Ruffed Grouse—Once found in wooded areas throughout region, now restricted as a native species to Berrien County where it is an uncommon permanent resident.

Wild Turkey—Once native species became extirpated over a century ago. Restocked in various places, but status as established species is questionable.

Northern Bobwhite—Fairly common to uncommon permanent resident in southernmost tier of Chicago-region counties, but rare elsewhere. Prefers fields with brushy areas for nesting.

Yellow Rail—Very rare from mid-April through mid-May and from mid-September through mid-October along lakefront and in grassy wetlands.

King Rail—Uncommon to rare from late April through mid-September. Breeds on edges of large marshes but can show up anywhere during migration. Has summered at Goose Lake Prairie, Dead Stick Pond, George Lake, Illinois Beach State Park, and Bong State Recreation Area.

Virginia Rail—Fairly common to uncommon from mid-April through mid-September. Breeds in marshes; during migration can be seen virtually anywhere.

Sora—Common to fairly common from early April through mid-October. Breeds in marshes; during migration can be seen virtually anywhere.

Common Moorhen—Uncommon from mid-April through late September in large marshes. Favored areas include Big Marsh, Des Plaines Wetlands Restoration Project, Cuba Road Marsh, and Haf's Pond.

American Coot—Abundant from late March through mid-May and late September through mid-November on lakes (shallow and deep) and sloughs. Less common as a migrant on Lake Michigan. Nests in marshes where it is fairly common to uncommon.

Sandhill Crane—Fairly common from mid-March through late April and from late October through early November. Can be seen migrating anywhere. Biggest concentration at Jasper-Pulaski Fish and Game Area during spring and fall. Uncommon nesting species in large marshes in Illinois (McHenry and Lake Counties) and Wisconsin (Walworth, Kenosha, and Racine Counties). Nesting sites include Chain O'Lakes State Park, Illinois Beach State Park, Tichigan Marsh, and Glacial Park.

Black-bellied Plover—Fairly common to uncommon from early May through early June and from early August through late October on Lake Michigan beaches, mud flats, flooded fields, and sewage ponds.

American Golden-Plover—Fairly common to uncommon from late March through mid-May and from mid-August through mid-October on flooded fields, sod farms, plowed fields, sewage ponds, and Lake Michigan beaches. Most common during spring in Newton County fields.

Semipalmated Plover—Fairly common to uncommon during May and from late July through late September on flooded fields, mudflats, sewage ponds, and Lake Michigan beaches.

Piping Plover—Rare from late April through mid-May and from late July through late September on Lake Michigan beaches.

Killdeer—Common from mid-March through late October. The most common local shorebird, nesting on short grass or gravel. During migration, prefers fields, mudflats, sewage ponds, sod farms, Lake Michigan beaches, but can be found almost anywhere.

American Avocet—Rare from mid-April through mid-May and from early July through late October. Most often seen along shoreline of Lake Michigan.

Greater Yellowlegs—Fairly common to uncommon from mid-March through mid-May and from early August through early November on mudflats, flooded fields, sewage ponds, and less frequently, Lake Michigan.

Lesser Yellowlegs—Common to fairly common from mid-April through late May and from early July through early October in same habitat as Greater Yellowlegs.

Solitary Sandpiper—Fairly common to uncommon from late April through mid-May and from early July through early September on mudflats (including the edges of rivers), marshes, and sewage ponds.

Willet—Rare from mid-April through mid-May and from mid-July through mid-September. Prefers shoreline of Lake Michigan but also found in sewage ponds and mudflats.

Spotted Sandpiper—Fairly common from mid-April through early October on muddy banks of rivers and lakes. During migration also found on sandy and rocky areas along Lake Michigan shoreline.

Upland Sandpiper—Uncommon to rare from mid-April through late August in areas of short grass including airports, pastures, and sod farms. Biggest nesting population is at Midewin National Grassland.

Whimbrel—Rare in May and from late July through late September. Prefers shoreline of Lake Michigan.

Hudsonian Godwit—Very rare in May and from late August through mid-October. Prefers Lake Calumet wetlands, flooded fields, and Lake Michigan shoreline.

Marbled Godwit—Rare in May and from late July through early September. Prefers Lake Michigan shoreline and Lake Calumet wetlands.

Ruddy Turnstone—Fairly common to uncommon from early May through early June and from early August through late September, mostly along Lake Michigan shoreline.

Red Knot—Rare from mid-May through early June and from mid-August through late September. Most sightings from Lake Michigan shoreline, with fewer number from Lake Calumet wetlands.

Sanderling—Fairly common to uncommon from early May through mid-June and from late July through early November along Lake Michigan shoreline.

Semipalmated Sandpiper—Fairly common to uncommon from early May through early June and common to fairly common from early July through late September. Prefers mudflats, wet fields, sewage ponds, and Lake Michigan shoreline.

Western Sandpiper—Very rare in May and uncommon to rare from early July through end of September. Similar habitat to that used by Semipalmated Sandpiper.

Least Sandpiper—Common to fairly common from early May through early June and from late July through late September. Similar habitat to that used by Semipalmated Sandpiper.

White-rumped Sandpiper—Uncommon from mid-May through mid-June and rare from late July through early November. Prefers mudflats, wet fields, and sewage ponds.

Baird's Sandpiper—Uncommon to rare from late April through late May and uncommon from early August through early September. Prefers Lake Michigan beaches, mudflats, sod farms, wet fields, and sewage ponds.

Pectoral Sandpiper—Fairly common from early April through mid-May and from mid-July through mid-November. Habitat similar to Baird's Sandpiper.

Purple Sandpiper—Rare from late October through early January on rocky areas along Lake Michigan.

Dunlin—Fairly common to uncommon from mid-May through mid-June and from early October through late November along Lake Michigan shoreline.

Stilt Sandpiper—Rare in May and uncommon from early July through early September. Prefers mud flats, flooded fields, and sewage ponds.

Buff-breasted Sandpiper—Uncommon to rare from late July through late September. Prefers sod farms, but also found at sewage ponds and along Lake Michigan beaches.

Ruff—Very rare from late April through mid-May and from early July through mid-August. Prefers mud flats, flooded fields, and sewage ponds.

Short-billed Dowitcher—Uncommon in May and fairly common to uncommon from mid-July through mid-September. Prefers mudflats and sewage ponds.

Long-billed Dowitcher—Uncommon to rare from late August through late October. Prefers mudflats and sewage ponds.

Common Snipe—Fairly common to uncommon in April and from early September through late October. Prefers mudflats with some vegetation to provide cover.

American Woodcock—Fairly common to uncommon from early March through mid-October. Prefers shrubby areas or open woods near wetlands. Difficult to find except during early spring when birds perform nuptial displays at dawn and dusk.

Wilson's Phalarope—Uncommon to rare from late April through early September. Prefers mudflats, shallow marshes, and sewage ponds.

Red-necked Phalarope—Very rare in May and rare from mid-August through late September. Prefers Lake Michigan shoreline, mudflats, and sewage ponds.

Red Phalarope—Rare to very rare from early October through late November along Lake Michigan shoreline.

Pomarine Jaeger—Very rare in November on Lake Michigan. NOTE: Because all Jaegers are so similar to one another and are often seen at great distances under adverse weather conditions, most jaegers are not identified as to species. The single best spot for all three jaegers is Miller Beach.

Parasitic Jaeger—Rare from early September through late November on Lake Michigan.

Long-tailed Jaeger—Very rare from late August through early October on Lake Michigan. Virtually all records are from Miller Beach.

Laughing Gull—Very rare from mid-April through mid-October along Lake Michigan shoreline.

Franklin's Gull—Uncommon to rare from early April through mid-November, most often along Lake Michigan shoreline.

Little Gull—Very rare from mid-April to mid-May and rare from late July through late December along Lake Michigan shoreline. Fraternizes with Bonaparte's Gull.

Bonaparte's Gull—Common from early April through early May and from late October through early December. Prefers Lake Michigan, but also occurs in flooded fields and smaller lakes, particularly in spring.

Ring-billed Gull—Abundant to common year-round resident throughout area. Largest nesting colony is at Lake Calumet.

Herring Gull—Common to fairly common year-round resident. Prefers garbage dumps and shore of Lake Michigan. Largest nesting colony is at Lake Calumet.

Thayer's Gull—Uncommon to rare from late November through early May along Lake Michigan and in garbage dumps.

Iceland Gull—Rare to very rare from mid-December through early May. Prefers Lake Michigan and garbage dumps, particularly those in Lake Calumet area.

Lesser Black-backed Gull—Rare from early November through early May. Prefers Lake Michigan and garbage dumps, particularly those in Lake Calumet area.

Glaucous Gull—Uncommon to rare from late November through early May. Prefers Lake Michigan and garbage dumps, particularly those in Lake Calumet areas.

Great Black-backed Gull—Rare from early September through late March. Prefers Lake Michigan and garbage dumps, particularly those in Lake Calumet area.

Sabine's Gull—Rare to very rare from mid-September through mid-November on Lake Michigan. Best location is Miller Beach on days with northerly winds.

Black-legged Kittiwake—Rare from late October through early December on Lake Michigan.

Caspian Tern—Fairly common to uncommon from late April through late September. Prefers Lake Michigan, but also sometimes found on inland lakes, particularly Lake Calumet.

Common Tern—Fairly common in May and from early August through late September on Lake Michigan. Breeds in the vicinity of Waukegan Beach.

Forster's Tern—Fairly common from mid-April through mid-May and from late August through early October. Prefers Lake Michigan, but also occurs inland, particularly in the Chain O'Lakes where it usually nests.

Black Tern—Uncommon from early May through late September on Lake Michigan, inland lakes, and sloughs during migration. Breeds in ever-decreasing number of marshes, particularly in northwestern part of the region.

Rock Dove—Introduced permanent resident now abundant throughout the region. Greatest numbers in highly urbanized areas.

Eurasian Collared-Dove—Introduced species that is increasing in numbers and range.

Mourning Dove—Common to fairly common throughout the year in all but most urbanized of areas.

Monk Parakeet—Introduced permanent resident whose status as an established species is unclear. Best-known colony is in Jackson Park and adjacent neighborhoods.

Black-billed Cuckoo—Uncommon to rare from early May through late September.

During migration found in lakefront parks and other locations with some woody vegetation. Breeds in woods and along edges of woods. Both species of cuckoos are attracted to tent caterpillars.

Yellow-billed Cuckoo—Uncommon from mid-May through late September. Similar habitat to Black-billed Cuckoo. Abundance of both species may vary greatly from year to year.

Eastern Screech-Owl—Fairly common to uncommon permanent resident in woods throughout region. Rarely detected except when it responds to an imitation or tape of its call at night.

Great Horned Owl—Uncommon permanent resident in woods throughout region.

Snowy Owl—Uncommon to rare from mid-November through mid-March. Varies tremendously in number from year to year—some winters virtually absent. Prefers Lake Michigan shoreline, particularly harbors. Single most reliable location is Meigs Field.

Barred Owl—Uncommon permanent resident in large bottomland woods. Responds well to tape or imitation of its call. Good locations include Indiana Dunes State Park, Warren Woods, and refuges in the Kankakee River basin.

Long-eared Owl—Uncommon to rare from mid-November through early April. During migration occurs on lakefront, often in fairly exposed locations. During winter, roosts in dense stands of evergreens and, less often, deciduous thickets.

Short-eared Owl—Uncommon from early October through late April. During migration, may be seen during the day flying along lakeshore. In winter, roosts in areas of large grasslands such as Bong State Recreation Area, Glacial Park, Pratt's Wayne Woods, and Fermilab.

Northern Saw-whet Owl—Rare from late October through late March in dense thickets or coniferous plantings. During migration, occurs on lakefront.

Common Nighthawk—Fairly common to common from mid-May through early September. During summer, most often found in urbanized areas where birds nest on flat roofs. Often heard at night. Forms huge flocks at dusk from late August through early September.

Chuck-wills-widow—Very rare from late April through early July. Best location is Mt. Baldy, where the bird is often heard at night.

Whip-poor-will—Fairly common to uncommon from mid-April through mid-September. Migrates along Lake Michigan, but far more abundant as a breeder in the southern Chicago-region counties where frequently heard calling at night in wooded areas. Favored locations include Illinois Beach State Park, Long John Slough, Cowles Bog, Inland Marsh, Willow Slough, Kankakee Fish and Game Area, and LaSalle Fish and Game Area.

Chimney Swift—Abundant to common from early May through early October. Seen hunting insects in the air.

Ruby-throated Hummingbird—Uncommon from early May through early September. Large numbers have been seen migrating along Lake Michigan and through the Indiana Dunes. Breeds in open woods, wood edges, shrubby areas, and other places supporting flowering plants throughout the nesting season.

Belted Kingfisher—Uncommon from late March through mid-October in vicinity of rivers, lakes, and sloughs.

Red-headed Woodpecker—Fairly common from mid-April through mid-October in open woods, especially where dead trees remain standing. Often winters in years with good acorns.

Red-bellied Woodpecker—Fairly common permanent resident in mature woods. Has been steadily increasing over the years.

Yellow-bellied Sapsucker—Fairly common to uncommon in April and from mid-September through mid-October.

Downy Woodpecker—Common to fairly common permanent resident in any habitat with trees.

Hairy Woodpecker—Fairly common to uncommon permanent resident in sites with mature trees.

Northern Flicker—Common to fairly common from late March through late October. Often migrates along Lake Michigan shoreline in substantial numbers; breeds in open woods.

Pileated Woodpecker—Rare permanent resident in mature woods. Favored areas include Ryerson Woods, Columbia Woods, Warren Woods, LaSalle Fish and Game Area, and Indiana Dunes State Park.

Olive-sided Flycatcher—Uncommon to rare from mid-May through early June and from mid-August through early September. Appears on lakefront and in woods and wooded edges with dead snags.

Eastern Wood-Pewee—Common to fairly common from early May through late September. Appears as a migrant along lakefront as well as inland areas. Breeds in mature woods.

Yellow-bellied Flycatcher—Uncommon from mid-May through early June and uncommon to rare from mid-August through mid-September in open woods and along lakefront.

Acadian Flycatcher—Uncommon from mid-May through early August in mature lowland woods. Favored locations include Wright Woods, Indiana Dunes State Park, Warren Woods, and the Heron Rookery. NOTE: The Acadian, Alder, Willow, and Least Flycatchers are all difficult to identify when not singing. This produces at least two consequences: (1) it is best not to wait much beyond June to look for them, and (2) their status in the fall is poorly defined and therefore not discussed.

Alder Flycatcher—Uncommon to fairly common from mid-May through mid-June in bogs and shrubby areas on the edge of marshes. Summering locations include Rochester area, Grant Preserve, Spring Creek Valley Forest Preserve, Volo Bog, and Beverly Shores.

Willow Flycatcher—Fairly common from mid-May through early August in shrubby areas near water.

Least Flycatcher—Common during May and uncommon to rare nester in open woods and aspens on edges of interdunal marshes.

Eastern Phoebe—Fairly common to uncommon from late March through early October. Can appear almost anywhere in migration; prefers to nest in open woods or woodland edges near water.

Great Crested Flycatcher—Fairly common to uncommon from early May through early September. Widespread in migration; prefers to nest in woods with some mature trees.

Eastern Kingbird—Common from early May through mid-September. Often migrates along Lake Michigan. Breeds in old fields and areas with scattered trees, particularly near water.

Northern Shrike—Uncommon to rare from mid-October through early March in open areas with trees, fences, posts, or other perches from which it can hunt. Abundance highly variable from year to year.

Loggerhead Shrike—Rare from mid-March through late September in similar habitat to Northern Shrike. Most common in southern Chicago-region counties. Largest local population at Midewin National Grassland.

White-eyed Vireo—Uncommon to rare from early May through late August. Breeds in scrubby areas particularly near water. More common in southern Chicago-region counties. Favored breeding sites include Beverly Shores, Bachelor Grove, LeRoy Oakes Forest Preserve, Dresden Lock and Dam, and Des Plaines Conservation Area.

Bell's Vireo—Rare from early May through early August in scrubby areas. Most common in southern Chicago-region counties. Favored sites include Goose Lake Prairie, Des Plaines Conservation Area, LaSalle Fish and Game Area, Willow Slough Fish and Game Area, and Jasper-Pulaski Fish and Game Area.

Blue-headed Vireo—Uncommon to rare in May and from early September through early October along lakefront and other areas with trees.

Yellow-throated Vireo—Uncommon from early May through early September. Breeds in mature woods. Favored breeding sites include Indiana Dunes State Park, woods along Des Plaines River, and woods along the Kankakee River.

Warbling Vireo—Common to fairly common from early May through mid-September. Prefers wooded areas near water.

Philadelphia Vireo—Uncommon in May and September along lakefront and other areas with trees.

Red-eyed Vireo—Common to fairly common from early May through mid-September. Breeds in woods with mature trees. In migration, along lakefront and other areas with trees or shrubs.

Blue Jay—Common to fairly common permanent resident in wide variety of habitats from woods to city neighborhoods. Many migrate along Lake Michigan.

American Crow—Abundant permanent resident in any habitat with trees, including city neighborhoods. Often gathers in large winter roosts.

Horned Lark—Fairly common to uncommon permanent resident in large, open areas with sparse vegetation.

Purple Martin—Common to fairly common from mid-April through mid-September. Prefers open areas near human habitation, particularly where martin houses are provided. Large numbers gather on lakefront in late summer prior to southward migration.

Tree Swallow—Common to fairly common from early April through early October. Most often seen feeding on insects over water. Nests in boxes and dead trees near water.

Northern Rough-winged Swallow—Fairly common to uncommon from mid-April through mid-September. Most often seen feeding on insects over water.

Bank Swallow—Fairly common to uncommon from early March through late August. Most often seen feeding on insects over water. Many migrate along Lake Michigan.

Cliff Swallow—Uncommon from late April through mid-September. Migrates in large numbers along Lake Michigan, particularly in fall. Nests locally with increasing frequency, often utilizing bridges.

Barn Swallow—Common from mid-April through mid-September. Many migrate along Lake Michigan. Most common nesting swallow.

Black-capped Chickadee—Abundant to common permanent resident in all wooded areas.

Tufted Titmouse—Fairly common to uncommon permanent resident in woods. Most common in southern Chicago region and Indiana Dunes.

Red-breasted Nuthatch—Uncommon from early September through early May. Prefers evergreens in winter, but any trees will do during migration. Numbers vary greatly from year to year.

White-breasted Nuthatch—Fairly common permanent resident in mature woods or even suburban areas with large trees.

Brown Creeper—Fairly common in April and from late September though early November. During migration is found most anywhere with trees. Uncommon during winter when it prefers riparian woods and evergreens.

Carolina Wren—Uncommon to rare permanent resident in ravines and lowland woods. More common in southern and eastern Chicago region. Severe winters reduce populations significantly.

House Wren—Common to fairly common from early May through mid-September in wooded areas with thick shrubby understory. Uses boxes or tree cavities for nesting sites, often in proximity to people.

Winter Wren—Fairly common to uncommon in April and fairly common in October. Rare in winter. Prefers dense cover, either in inland woods near water or along the lakefront.

Sedge Wren—Fairly common to uncommon from mid-May through mid-September. Often seen in migration along lakefront. Breeds in wetlands dominated by grasses and sedges. One of the few birds that thrives in solid stands of Reed Canary Grass.

Marsh Wren—Fairly common to uncommon from mid-May through late September. Often seen in migration along lakefront. Breeds in cattail marshes.

Golden-crowned Kinglet—Fairly common from mid-March through late April and from early October through early November. Found almost anywhere in migration. Uncommon to rare in winter, when it prefers evergreens and lowland woods.

Ruby-crowned Kinglet—Fairly common from early April through mid-May and from early September through late October. Found almost anywhere in migration.

Blue-gray Gnatcatcher—Fairly common to uncommon from mid-April though late July. Breeds in mature woods and sand savannas; found in migration most anywhere there are trees. Increasing in abundance.

Eastern Bluebird—Fairly common to uncommon from mid-March through early November in areas with scattered trees. Have increased with placement of nesting boxes. Often seen migrating in spring at hawk watching sites in Indiana Dunes.

Veery—Fairly common in May and uncommon to rare from early June through late July. Seen on lakefront and inland areas during migration where there is sufficient vegetation for shelter. Breeds in mature woods with dense understory.

Gray-cheeked Thrush—Uncommon in May and September. Found along lakefront and other areas with sufficient trees and understory.

Swainson's Thrush—Common in May and from late August through late September. Same habitat as Gray-cheeked Thrush.

Hermit Thrush—Fairly common in April and from late September through late October. Same habitat as Gray-cheeked Thrush. Rare in winter in lowland woods and evergreens.

Wood Thrush—Fairly common from early May through late August. Breeds in mature woods. During migration found along lakefront and other areas with sufficient trees and understory.

American Robin—Abundant to common from early March through mid-October. Uncommon in winter. Greater numbers present in winter due, at least in part, to spread of non-native fruit-bearing shrubs.

Varied Thrush—Very rare from early December through late March. Most often detected at feeders.

Gray Catbird—Common from early May through early October. Breeds where there is shrubby understory: in woods, wood edges, old fields, and even backyards. Can be found almost anywhere during migration, particularly along lakefront.

Northern Mockingbird—Uncommon to rare from early April though late August. Breeds in scrubby areas. Most common in southern tier of Chicago-area counties.

Brown Thrasher—Fairly common to uncommon from mid-April through mid-October. Breeds in scrubby areas but can be found during migration nearly anywhere.

European Starling—Introduced permanent resident, now abundant throughout the region.

American Pipit—Uncommon to rare from mid-March through late May and from mid-September through late October in muddy fields, sod farms, mowed areas on lakefront, and beaches.

Bohemian Waxwing—Very rare from mid-November through late March in areas with fruit-bearing trees and shrubs.

Cedar Waxwing—Fairly common to common permanent resident almost anywhere except heavily urbanized areas. Large numbers migrate along lakefront. In winter, most often found foraging on fruit.

Blue-winged Warbler—Fairly common to uncommon from early May through late August. Nests in scrubby areas and woodland edges throughout region. NOTE: During migration, warblers can show up in virtually any terrestrial environment, from the best forests to the Loop in downtown Chicago. Thus, this discussion on warblers generally omits migration habitat.

Golden-winged Warbler—Uncommon in May and from mid-August through mid-September.

Tennessee Warbler—Common in last three weeks of May and from mid-August through late September.

Orange-crowned Warbler—Uncommon to rare from mid-April through mid-May and from mid-September through late October.

Nashville Warbler—Common to uncommon from late April through late May and from late August through early October.

Northern Parula—Uncommon in May and uncommon to rare in September. During fall, mostly seen on lakefront.

Yellow Warbler—Common to fairly common from late April through mid-August. Most common nesting warbler, preferring thickets near water.

Chestnut-sided Warbler—Common in May and fairly common from late August through late September. Rare nester in scrubby areas on edge of woods.

Magnolia Warbler—Common in May and from late August through late September.

Cape May Warbler—Fairly common to uncommon in May and from late August through mid-September.

Black-throated Blue Warbler—Uncommon to rare in May and from late August through early October.

Yellow-rumped Warbler—Abundant from mid-April through mid-May and common from mid-September through late October. Warbler most likely to winter, during which season it prefers evergreens, wooded areas with open water, and fruit-bearing shrubbery.

Black-throated Green Warbler—Common in May and from late August through early October.

Blackburnian Warbler—Fairly common to uncommon in May and from late August through mid-September.

Yellow-throated Warbler—Rare from mid-April through mid-June. Most common in Indiana counties and Berrien County where it nests in mature riparian woods.

Favored nesting sites include Heron Rookery, Kankakee Fish and Game Area, and Forest Lawn/Lakeside Roads.

Pine Warbler—Rare in late April through mid-May and in September.

Prairie Warbler—Rare from late April through mid-September. For many years an isolated nesting colony has persisted at Indiana Dunes State Park where they can be heard in May and June.

Palm Warbler—Common from late April through mid-May and from early September through mid-October.

Bay-breasted Warbler—Common to fairly common in May and from late August through end of September.

Blackpoll Warbler—Common to fairly common in May and from late August through late September.

Cerulean Warbler—Uncommon to rare from early May through late August. Favored summering locations include Oak Point, Indiana Dunes State Park (the best), Heron Rookery, Moraine State Nature Preserve, and Warren Woods.

Black-and-white Warbler—Fairly common in May and from late August through late September.

American Redstart—Common from mid-May through early June and from mid-August through mid-September. Uncommon to rare in summer when it nests in bottomland woods, particularly in Indiana. Best summering locations are Heron Rookery and Indiana Dunes State Park.

Prothonotary Warbler—Uncommon to rare from early May through early September. Nests in riparian woods. Favored summering locations include Kankakee Fish and Game Area, LaSalle Fish and Game Area, Indiana Dunes State Park, Sarrett Nature Center, and McHenry Dam.

Worm-eating Warbler—Rare from late April through mid-May. Birds have summered at Grand Mere. Rarely seen in fall.

Ovenbird—Fairly common to uncommon from early May through late September. Nests throughout region in mature woods.

Northern Waterthrush—Fairly common from mid-April through mid-May and from late August through late September.

Louisiana Waterthrush—Rare from mid-April through late July. Favored summering locations include Indiana Dunes State Park, Heron Rookery, Warren Woods, and McClaughry Springs Woods.

Kentucky Warbler—Rare to very rare from early May through late July. Breeds in mature woods with well-developed understory. Restoration activities at Swallow Cliff destroyed last reliable summering site. Other locations where bird has summered include Indiana Dunes State Park, Ryerson Woods, Pilcher Park, and Green Lake Woods.

Connecticut Warbler—Uncommon to rare in the last two weeks of May and in September.

Mourning Warbler—Uncommon during the last three weeks of May and from late August through mid-September.

Common Yellowthroat—Common to fairly common from early May through late September. Second most common nesting warbler. Nests in wet scrubby areas often on edges of marsh.

Hooded Warbler—Rare from late April through early September. Summering locations include Indiana Dunes State Park, Ly-co-ki-we Trail, Warren Dunes, Warren Woods, and Ryerson Woods.

Wilson's Warbler—Fairly common in May and from late August through mid-September.

Canada Warbler—Uncommon in last three weeks of May and from mid-August through mid-September. Rare in summer, when it prefers low woods with well-developed understory. Two favored summering locations are Indiana Dunes State Park and Grand Mere.

Yellow-breasted Chat—Uncommon from early May through mid-August. Nests in scrubby areas. Favored summering locations include Beverly Shores, Cowles Bog, Willow Slough Fish and Game Area, LaSalle Fish and Game Area, and Des Plaines Conservation Area.

Summer Tanager—Rare from late April through late July. It regularly summers in sandy open woods. Has summered at Indiana Dunes State Park, Willow Slough Fish and Game Area, and Grand Mere.

Scarlet Tanager—Fairly common to uncommon from early May through late September. Breeds in mature woods, and migrates wherever there are large trees.

Eastern Towhee—Common to fairly common from late March through mid-October. During migration can be found most anywhere. Breeds most commonly in the southern Chicago region, where it prefers scrubby areas.

American Tree Sparrow—Common from late October through mid-April in fallow cropland, scrubby areas, and marshes.

Chipping Sparrow—Common to fairly common from mid-April through late September in areas combining short grass with trees and shrubs. Favors cemeteries, golf courses, and suburban neighborhoods.

Clay-colored Sparrow—Rare in May and from mid-September through mid-October. Most often found in lakefront parks.

Field Sparrow—Common to fairly common from early April through mid-October. Breeds in open areas with scrub.

Vesper Sparrow—Uncommon from early April through late September. Breeds in fallow fields and sandy prairie. Decreasing with loss of farmland. Best way to find the bird is to cruise rural roads in May and June and listen for its song.

Lark Sparrow—Rare from late April through late August. Breeds in sandy scrub of Morris-Kankakee Basin. Favored sites include Braidwood Dunes and Savanna, Pembroke Township, and Willow Slough (best).

Savannah Sparrow—Common to fairly common from early April through late October in areas of extensive grass.

Grasshopper Sparrow—Fairly common to uncommon from late April through late August in areas of extensive grass.

Henslow's Sparrow—Uncommon to rare from late April through late August. During migration often seen in lakefront parks. Breeds in areas of mixed grass and forbs. Favored nesting sites include Goose Lake Prairie (best), Newton County Fields, Bong State Recreation Area, Orland Hills Forest Preserve, and Vollmer Road Wildlife Marsh.

Le Conte's Sparrow—Rare from early April through early May and from early September through mid-October. Most often found in lakefront parks amidst grass.

Nelson's Sharp-tailed Sparrow—Rare from late April through late May and from early September through mid-October in lakefront parks and wet grassy areas.

Fox Sparrow—Fairly common to uncommon from late March through mid-April and from early October through early November in weedy fields, wood edges, and lakefront parks.

Song Sparrow—Common to fairly common permanent resident (sometimes uncommon in winter) in a wide range of open habitats.

Lincoln's Sparrow—Uncommon in May and from mid-September through mid-October along lakefront and in wet woods and scrub.

Swamp Sparrow—Common to fairly common from early April through late October. Breeds in marshes, migrates along lakefront and in scrubby fields, and winters (uncommon to rare) in marsh or scrub near open water.

White-throated Sparrow—Common from mid-April through mid-May and from early September through late October in woodlands, woodland edges, and lakefront. Uncommon to rare in winter in same habitat.

Harris's Sparrow—Very rare from late April through mid-May and rare from mid-September through mid-October. Most often seen along the lakefront.

White-crowned Sparrow—Fairly common from mid-April through mid-May and from mid-September through late October in fallow fields, scrub, and lakefront.

Dark-eyed Junco—Abundant to common from mid-September through mid-April in scrubby fields, suburban neighborhoods, and other semi-open habitat and along wood edges and lakefront.

Lapland Longspur—Uncommon from late September through mid-April in sparsely vegetated fields and along lakefront. Prefers muddy fields with soybean stubble.

Smith's Longspur—Rare from early March through early May in sparsely vegetated fields. Most common in Newton County. Prefers fields with corn stubble and foxtail over heavy crop litter.

Snow Bunting—Fairly common to uncommon from late October through late March in sparsely vegetated fields and along the lakefront.

Northern Cardinal—Abundant permanent resident wherever there are trees or shrubs.

Rose-breasted Grosbeak—Fairly common from early May through mid-September in woods.

Blue Grosbeak—Rare from mid-May through late July. Most common in southern Chicago-region counties where it has summered in sandy scrub at Willow Slough and Beaver Lake Prairie Chicken Refuge.

Indigo Bunting—Common to fairly common from mid-May through late September in woodland edges with dense cover.

Dickcissel—Uncommon from mid-May through mid-August in abandoned cropland. More common south of Chicago. Varies in abundance from year to year.

Bobolink—Fairly common to uncommon from early May through early September. Migrates along lakeshore and breeds in grassy fields. Species has been declining steadily.

Red-winged Blackbird—Abundant from early March through late October in open areas ranging from marshes to fields. Rare in winter, except where birds roost in large flocks.

Eastern Meadowlark—Fairly common from mid-March through mid-October in grasslands.

Western Meadowlark—Uncommon to rare from early March to mid-October. Prefers fields slightly higher and less vegetated than those favored by Eastern Meadowlark. Favored breeding sites include Illinois Beach State Park, Kettle Moraine State Forest, Hampshire Forest Preserve, Des Plaines Conservation Area, and McHenry County sod farms.

Yellow-headed Blackbird—Uncommon from mid-April through mid-September in marshes with open water. Favored locations include Des Plaines Wetland Demonstration Project, Hegewisch Marsh, Nippersink Marsh, Moraine Hills State Park, and Bong State Recreation Area.

Rusty Blackbird—Fairly common to uncommon from late March through late April and from early October through early November in wet woods.

Brewer's Blackbird—Rare permanent resident. Breeds in sparsely vegetated grasslands. Often winters in vicinity of livestock. Two reliable sites are Illinois Beach State Park (South Unit) during May and June and Shamrock Turf Nurseries in early fall.

Common Grackle—Abundant in all habitats and in all seasons except winter. Rare in winter except where large roosts are present.

Brown-headed Cowbird—Common to fairly common from mid-March through mid-November in almost all open and semi-open areas. Less common in late summer after nesting, and rare in winter except where part of large blackbird flocks.

Orchard Oriole—Uncommon to rare from late April through late July in open woods. More common in southern Chicago region.

Baltimore Oriole—Fairly common from early May to early September in open woods with mature trees.

Pine Grosbeak—Very rare from mid-November through early March at feeders, evergreens, and fruit-bearing trees. Not present every year, but periodically appears in small numbers.

Purple Finch—Uncommon from mid-September through late April at feeders and in woods. Has decreased markedly over the years.

House Finch—Introduced species that has become a common permanent resident. Most common in developed areas with trees and shrubs.

Red Crossbill—Uncommon to rare from early October through mid-March in areas with extensive pines. Numbers highly variable from year to year.

White-winged Crossbill—Rare to very rare from early November through late March in areas with extensive evergreens. During invasion years, fair numbers may appear. Two of the best locations to find this bird are Lyons Woods and the Morton Arboretum.

Common Redpoll—Rare from early November through mid-March at feeders, in seedy fields, and areas with evergreens. During invasion years, large numbers may appear.

Pine Siskin—Uncommon to rare from mid-October through early April at thistle-seed feeders and areas with evergreens, alders, and/or birches.

American Goldfinch—Common to fairly common permanent resident in seedy fields and at feeders and scrub.

Evening Grosbeak—Rare from early November through early April at feeders and in areas with evergreens, birch, fruit trees, and other food-bearing trees. During invasion years, substantial numbers may appear.

House Sparrow—Introduced permanent resident that is abundant throughout the region.

Organizations and Resources

National

American Birding Association
P.O. Box 6599, Colorado Springs, CO 80934 (800-850-2473)

National Audubon Society
(NAS), P.O. Box 51003, Boulder, CO 80323-1003 (800-274-4201)

National Biological Survey
Bird Banding Laboratory, 12100 Beech Forest Rd., Laurel, MD 20708-4037

Illinois

Chicago Academy of Sciences
2340 N. Cannon Dr., Chicago, IL 60614 (773-549-0606)

Chicago Audubon Society
(NAS chapter), 5801 N. Pulaski Rd., Chicago, IL 60646 (773-539-6793)

Chicago Ornithological Society
c/o 4016 N. Clarendon Ave. #3N, Chicago, IL 60613

Chicago Park District
425 E. McFetridge Dr., Chicago, IL 60502 (312-747-2200)

DuPage Audubon Society
(NAS chapter), P.O. Box 1201, Wheaton, IL 60189-1201

DuPage Birding Club
405 Washington St., Elmhurst, IL 60126 (630-406-8111)

Evanston North Shore Bird Club
P.O. Box 1313, Evanston, IL 60204

Field Museum of Natural History
Roosevelt Rd. at Lake Shore Dr., Chicago, IL 60605 (312-922-9410)

Forest Preserve District of Cook County
536 North Harlem Ave., River Forest, IL 60305 (800-870-3666)

Forest Preserve District of DuPage County
120 E. Liberty Dr., Wheaton, IL 60187 (630-933-7200)

Forest Preserve District of Kane County
719 Batavia Ave., Bldg. G, Geneva, IL 60134 (630-232-5980)

Forest Preserve District of Kendall County
111 West Fox St., Yorkville, IL 60560-1498 (630-553-4131)

Forest Preserve District of Lake County
2000 North Milwaukee Ave., Libertyville IL 60048 (847-367-6640)

Forest Preserve District of Will County
22606 South Cherry Hill Rd., Joliet, IL 60433 (815-727-8700)

Illinois Audubon Society
(IAS) Headquarters, P.O. Box 2418, Danville, IL 61834 (217-446-5085)

Illinois Audubon Society, Fort Dearborn Chapter
943 S. East Ave., Oak Park, IL 60304

Illinois Audubon Society, Kane County Chapter
P.O. Box 201, Batavia, IL 60510

Illinois Audubon SocietyLake-Cook Chapter
P.O. Box 254, Highland Park, IL 60035 (847-432-0706)

Illinois Audubon Society, McHenry County Chapter
P.O. Box 67, Woodstock, IL 60098 (815-338-0592)

Illinois Audubon Society, Park Ridge Chapter
P.O. Box 217, Park Ridge, IL 60068

Illinois Audubon Society, Will County Chapter
P.O. Box 3261, Joliet, IL 60434 (815-741-7277)

Illinois Ornithological Society
P.O. Box 931, Lake Forest, IL 60045-0931

Lake Forest Open Lands Association
272 Market Square, Suite #2726, Lake Forest, IL 60045 (847-234-3880)

McHenry County Conservation District
6512 Harts Rd., Ringwood, IL 60072 (815-678-4431)

Prairie Woods Audubon Society (NAS chapter)
P.O. Box 1065, Arlington Heights, IL 60006

Sand Ridge Audubon Society
15891 Paxton Ave., South Holland, IL 60473 (708-868-0606)

The Nature Conservancy
Illiniois Field Office, 8 S. Michigan Ave., Chicago, IL 60603 (312-346-8166)

Thorn Creek Audubon Society (NAS chapter)
P.O. Box 895, Park Forest, IL 60466

Indiana

Dunes-Calumet Audubon Society (NAS chapter)
P.O. Box 15, Valparaiso, IN 46383-0015

Friends of Gibson Woods
6201 Parrish Ave., Hammond, IN 46323 (219-844-3188)

Indiana Audubon Society
c/o Mary Gray Bird Sanctuary, 3499 S. Bird Sanctuary Rd., Connersville, IN 47331 (765-825-9788)

Pottawotamie Audubon Society (NAS chapter)
412 West 250S, LaPorte, IN 46350

South Bend Audubon Society (NAS chapter)
P.O. Box 581, Mishawaka, IN 46545 (219-243-8739)

The Nature Conservancy, Indiana Field Office
1330 West 38th St., Indianapolis, IN 46208 (317-923-7547)

Michigan

Berrien Birding Club
9228 Huckleberry Rd., Berrien Center, MI 49102 (616-471-2617)

Michigan Audubon Society (NAS chapter)
6011 W. St. Joseph Hwy., Lansing, MI 48917 (517-886-9144)

The Nature Conservancy, Michigan Field Office
2840 E. Grand River Ave., East Lansing, MI 48823 (517-332-1741)

Wisconsin

Hoy Nature Club
925 Lawn Ave., Racine, WI 53405

Kenosha County Parks
19600 75th St., P.O. Box 549, Bristol, WI 53104 (262-857-1869)

Lakeland Audubon Society (NAS chapter)
P.O. Box 473, Elkhorn, WI 53121

Racine County Parks
14200 Washington Ave., Sturtevant, WI 53177-1253 (262-886-8440)

The Nature Conservancy
Wisconsin Field Office, 633 West Main St., Madison, WI 53703 (608-251-8140)

Wisconsin Society for Ornithology
c/o W330 N8275 W. Shore Dr., Hartland, WI 53029-9732 (262-966-1072)

Nature Centers

THE FOLLOWING NATURE CENTERS are additions to those described in the text. Most of these have naturalists and public programs and can be a good source of information about birds. Bird feeders often attract interesting species, and the grounds offer good birding in a safe environment. Call to see what days and what hours each center is open to visitors.

In Cook County, Illinois, the following nature centers are in addition to those at Camp Sagawau, Crabtree, Little Red Schoolhouse, River Trail, and Sand Ridge:

North Park Village Nature Center, 5800 North Pulaski Rd., Chicago, IL 60646. Open grassy areas, woods and wetland on the grounds of the former Chicago Tuberculosis Sanitarium, the nature center is owned by the Chicago Park District and is the meeting place for the Chicago Audubon Society. Telephone— (773)744-5472

Emily Oaks Nature Center, 4650 Brummel St., Skokie, IL 60076. This 13-acre preserve is wooded and has a pond. It hosts birding activities on some Saturdays. Owned by the Skokie Park District. Telephone—(847)677-7001

Evanston Ecology Center, 2024 McCormick Blvd., Evanston, IL 60201. Located on 17 acres on the west bank of the North Shore Channel, the center is the location for meetings of the Evanston North Shore Bird Club. Telephone—(847)864-5181

Wildwood Nature Center, 2701 Sibley Ave., Park Ridge, IL 60068. This small center has a pond that attracts herons and waterfowl. Owned by the Park Ridge Recreation and Park District. Telephone—(847)692-3570

Stillman Nature Center, 33 W. Penny Rd., Barrington, IL 60010. 80 acres with cattail marsh, small lake, pine plantation and restored prairie. Owned by a private foundation, the nature center is open to the public on Sundays from 9:00 A.M. to 4:00 P.M. (1:00 to 4:00 from November through March) and open during the week for groups by reservation. Telephone—(847)428-6957

Spring Valley Nature Center, 1111 E. Schaumburg Rd., Schaumburg, IL 60194. This 135-acre preserve is owned by the Schaumburg Park District. It has 3.5 miles of trails through a variety of habitats. Telephone—(847)985-2100

Lake Katherine Nature Preserve, 7402 W. Lake Katherine Dr., Palos Heights, IL 60463. A 91-acre park with a 20-acre lake, wetlands, waterfall garden, and small prairie on the south side of the Cal Sag Channel west of Harlem. Owned by the City of Palos Heights. Telephone—(708)361-1873

Irons Oaks Nature Center, 2453 Vollmer Rd., Olympia Fields, IL 60461. A 37-acre preserve owned by the park districts of Homewood, Flossmoor, and Olympia Fields. Telephone—(708)481-2330

Lake County has nature centers at Illinois Beach State Park, Volo Bog, Reed-Turner Woodland, and at the following:

Heller Nature Center, 2821 Ridge Rd., Highland Park, IL 60035. Ninety-seven acres of upland hardwoods, mesic and wet prairie, remnant pine nursery, and pond west of Highway 41 on Ridge Road. The center is the location for meetings of the Lake-Cook Chapter of Illinois Audubon Society. Owned by the Park District of Highland Park. Telephone—(847)433-6901

Prairie Grass Nature Museum, 860 Hart Rd., Round Lake, IL 60073. Owned by the Round Lake Park District. Telephone—(847)546-8558

McHenry County has a nature center at Moraine State Park and at the following:

Prairieview Education Center, 2112 Behan Road, Crystal Lake, IL 60014. On 250 acres of uplands, restored prairie, and wetlands on the west bank of the Fox River, south of Rte. 176. Oned by the McHenry County Conservation District. Telephone—(815)479-5779

Veteran Acres Park, 330 N. Main St., Crystal Lake, IL 60014

DuPage County has Fullersburg Woods Environmental Education Center and the following:

Spring Brook Nature Center, 130 Forest Ave., Itasca, IL 60143. Eighty acres with marsh, woods, meadow, and stream. Raptors are rehabilitated at the center. Owned by the village of Itasca. Telephone—(630)773-5572

Lake View Nature Center, 17 W. 963 Hodges Rd., OakBrook Terrace, IL 60181. On 17 acres across from the Oak Brook shopping center, the center is owned by the Oak Brook Terrace Park District. Telephone—(630)941-8747

Willowbrook Wildlife Haven, 525 South Park Blvd., Glen Ellyn, IL 60137. This center rehabilitates native animals and has many on display. It also has a nature trail on its 50 acres and offers educational hikes and programs in various county preserves. Owned by the Forest Preserve District of DuPage County. Telephone—(630)942-6200

Kane County has Tekawitha Woods Nature Center and the following:

> **Red Oak Nature Center,** 2343 River St., Batavia, IL 60510. Forty acres of oak woods on the Fox River south of Batavia. Owned by the Fox Valley Park District. Telephone—(630)897-1808

> The **St. Charles Park District** has a naturalist and offers nature programs in several locations. Pottawatomie Community Center, 8 North Ave., St. Charles, IL 60174. Telephone—(630)513-3338

Will County has nature centers at Lake Renwick, Goodenow Grove, and Thorn Creek.

Kankakee County has a nature center at Kankakee State Park.

In Indiana, there are nature centers in Lake County at Gibson Woods, and in Porter County at Indiana State Park and at Indiana National Lakeshore Visitor Center.

In Michigan, Berrien County has nature centers at Love Creek Park, Fernwood Botanic Garden, and Sarett Nature Center.

In Wisconsin, Kenosha County has nature centers at Bong State Recreation Area and

> **Hawthorn Hollow Nature Sanctuary,** 880 Green Bay Rd., Kenosha, WI 53144. Forty acres with prairie, garden, and wooded bluffs near Petrifying Springs. Owned by a private foundation, the preserve is closed for the month of February. Telephone—(262)552-8196

Racine County has the following:

> **River Bend Nature Center,** 3600 N. Green Bay Rd., Racine, WI 53404. This small preserve with old fields, woods, and backwater area along the Root River is owned by the Racine YWCA. Telephone—(262)639-0930

Hotlines

Rare Bird Alerts

Chicago area, IL (847) 265-2118

DuPage County, IL (630) 406-8111

Rockford area, IL (815) 965-3095

State of Indiana (317) 259-0911

State of Michigan (616) 471-4919

Berrien County, MI (616) 982-8676 ext. 1

Racine County, WI (262) 835-0460

State of Wisconsin (414) 352-3857

Birding Websites

http://www.xnet.com/~ugeiser/Birds/IBET.html (Illinois)

http://www.uwgb.edu/wsonews.html (Wisconsin)

http://www.audbon.org/listserv/in-bird/html (Indiana)

http://warbler.med.umich.edu/Birders_FAQ.html (Southeast Michigan)

Christmas Bird Counts

Christmas Count Name	Sponsoring Organization
Barrington, IL	Praire Woods Audubon
Calumet City–Sand Ridge, IL	Sand Ridge Nature Center
Chicago Lakefront, IL	Evanston North Shore Bird Club
Chicago Urban, IL	Evanston North Shore Bird Club
DeKalb, IL	Illinois Ornithological Society
Evanston North Shore, IL	Evanston North Shore Bird Club
Fermilab-Batavia, IL	DuPage Birding Club
Joliet, IL	Pilcher Park
Lisle Arboretum, IL	Chicago Ornithological Society
McHenry County, IL	McHenry County Audubon
Morris-Wilmington, IL	Will County Audubon
Thorn Creek-Park Forest, IL	Thorn Creek Audubon
Waukegan, IL	Evanston North Shore Bird Club
Indiana Dunes National Lakeshore, IN	Chicago/Dunes-Calumet Audubon Society
Southeastern LaPorte County, IN	Indiana Audubon Society
Southern Lake County, IN	Indiana Audubon Society
Willow Slough–Iroquois Preserves, IN	Indiana Audubon Society
Berrien Springs, MI	Love Creek Nature Center
Coloma, MI	Love Creek Nature Center
New Buffalo, MI	Love Creek Nature Center
Niles, MI	Love Creek Nature Center
Burlington, WI	Wisconsin Society for Ornithology
Kenosha, WI	Wisconsin Society for Onithology
Lake Geneva, WI	Wisconsin Society for Onithology
Racine, WI	Wisconsin Society for Onithology

Rare Bird Documentation

To document unusual sighting(s) in Illinois,
provide the information called for on as many pages as necessary.

DOCUMENTATION FORM FOR UNUSUAL BIRD SIGHTINGS

Submitted to Illinois Ornithological Records Committee (I.O.R.C.)

This form is submitted as supporting documentation of (check all that apply):

❏ Unusual species   ❏ Unusual date;

❏ Unusual number   ❏ Unusual plumage;

❏ Unusual breeding record   ❏ Christmas Count record;

❏ Spring Count record   ❏ Breeding Census record;

❏ Other

1. Species: _____

2. Number of birds: Age / sex / plumage: _____

_____

3. Date(s): _____

4. Location (incl. County): _____

5. Observers: _____

     Your name: _____

     Phone: (   ) _____

     Mailing address: _____

     E-mail address: _____

     Others agreeing with identification: _____

_____

     Observers NOT agreeing with identification: _____

_____

6. Physical description (size, shape, proportions, details of both color and pattern on the head, back, breast, undertail, wings, and tail; coloration of bill, eye, legs and feed) of details *actually seen in the field*:

7. Description of behavior: _____

_____

8. Description of vocalizations: _____

_____

9. Description of immediate and surrounding habitat(s): _____

_____

10. Viewing conditions: _____

    Optical equipment (type, power): _____

    Distance (how measured): _____

    Time(s) of observation: _____

    Total time of observation: _____

    Weather (including larger weather patterns where relevant)/ sky conditions/
    relative position of sun: _____

11. Previous experience with this and similar species: _____

_____

12. Reasons for eliminating other similar species and/or hybrids: _____

_____

_____

13. Were photos obtained? _____ By whom? _____ Attached? _____

14. Books and illustrations consulted, and advice received. How did these influence this description?

_____

_____

15. How long after observation were field notes recorded? _____

16. How long after observation was this form completed?_____

17. Additional remarks: _____

_____

_____

_____

_____

_____

Signed _____ Date _____

Please mail original documentation immediately on completion to:

Avian Ecology Program, Natural Heritage Division/Department of Natural Resources, Springfield, IL 62701

Bibliography

American Ornithologist's Union Committee on Classification and Nomenclature. *Checklist of North American Birds*. 7th ed. Lawrence, Kan.: American Ornithologist's Union, 1998.

Bateman, Newton, and F. Wilcox. *Historical Encyclopedia of Illinois and History of Kane County*. Chicago: Munsell, 1904.

Binford, Laurence C. "Bird Finding Guide—The Chicago Botanic Garden." *Meadowlark* 2, no. 4 (1994): 131–34.

Biss, Richard. "Bird Finding Guide—Shorebirds at O'Hare Post Office Ponds." *Meadowlark* 4, no. 3 (1995): 101–4.

Bohlen, H. David. *The Birds of Illinois*. Bloomington: Indiana University Press, 1989.

Brock, Kenneth J. *Birds of the Indiana Dunes*. Michigan City, Ind.: Shirley Heinze Environmental Fund, 1997.

Clark, Charles T., and Margaret M. Nice. *William Dreuth's Study of Bird Migration in Lincoln Park,Chicago*. Chicago: The Chicago Academy of Sciences, 1950.

Daniel, Glenda. *Dune Country*. Rev. ed. Athens, Ohio: Swallow Press, 1984.

De Vore, Sheryl. "Morton Arboretum—A Year-Round Birding Mecca." *Wild Bird Magazine* (February 1989).

Engel, Ronald J. *Sacred Sands: The Struggle for Community in the Indiana Dunes*. Middletown, Conn.: Weslayan University Press, 1983.

Fawks, Elton. *Bird Finding in Illinois*. Downers Grove: Illinois Audubon Society, 1975.

Ford, Edward R. *Birds of the Chicago Region*. Chicago: The Chicago Academy of Sciences, 1956.

Fryxell, F. M. *The Physiography of the Region of Chicago*. Chicago: University of Chicago Press, 1927.

Greenberg, Joel. *A Natural History of the Chicago Region*. Baltimore, Md.: Johns Hopkins University Press, in press.

Heidorn, Randy R., William D. Glass, Daniel R. Ludwig, and Margaret A. R. Cole. *Northeastern Illinois Wetland Survey for Endangered and Threatened Birds*. Springfield: Illinois Department of Conservation, 1991.

Hickman, Scott. "The Des Plaines River Wetlands Demonstration Project." *Meadowlark* 1, no. 1 (1992): 9–13.

Indiana Division of Nature Preserves. *Directory of Indiana's Dedicated Nature Preserves*. Indianapolis: Indiana Department of Natural Resources, 1995.

Kania, Denis, and Peter Kasper. "Bird Finding Guide—Fermilab." *Meadowlark* 4, no. 2 (1995): 51–56.

Karnes, Jean, and Don McFall, eds. *A Directory of Illinois Nature Preserves*. Springfield: Illinois Nature Preserves Commission, 1995.

Keller, Charles E., Shirley A. Keller, and Timothy C. Keller. *Indiana Birds and*

Their Haunts. Bloomington: Indiana University Press, 1986.

Landing, James E. "Bird Finding Guide—Gull Birding in Illinois." *Meadowlark* 3, no. 2 (1994): 54–58.

Leopold, Aldo. *A Sand County Almanac*. San Francisco and New York: Sierra Club/Ballantine Books, 1970.

Marcisz, Walter J. "Bird Finding Guide—Lake Calumet." *Meadowlark* 3, no. 4 (1994): 133–35.

Mellin, Judy. "Bird Finding Guide—Poplar Creek Forest Preserve." *Meadowlark* 5, no. 2 (1996): 57–59.

Mlodinow, Steven. *Chicago Area Birds*. Chicago: Chicago Review Press, 1984.

Moskoff, William. "Birding Lake Forest's Open Lands." *Meadowlark* 2, no. 3 (1993): 96–98.

Mumford, Russell E., and Charles E. Keller. *The Birds of Indiana*. Bloomington: Indiana University Press, 1884.

Peattie, Donald Culross. *A Prairie Grove*. New York: Simon and Schuster, 1938.

Robbins, Samuel D., Jr. *Wisconsin Birdlife: Population & Distribution—Past & Present*. Madison: University of Wisconsin Press, 1991.

Sanders, Jeffrey, and Lynne Yaskot. *Top Birding Spots Near Chicago*. Rev. ed. Winnetka, Ill.: Evanston North Shore Bird Club, 1977.

Smith, Ellen Thorne. *Chicagoland Birds: Where and When to Find Them*. Rev. ed. Chicago: Field Museum of Natural History, 1972.

Swink, Floyd. *A Finding List For The Birds of Morton Arboretum*. Lisle, Ill.: The Morton Arboretum, 1976.

Temple, Stanley A., and John R. Cary. *Wisconsin Birds: A Seasonal and Geographical Guide*. Madison: University of Wisconsin Press, 1997.

Tessen, Daryl D. *Wisconsin's Favorite Bird Haunts*. Rev. ed. DePere: The Wisconsin Society for Ornithology, 1989.

White, John, and Michael Madny. *Illinois Natural Areas Inventory Technical Report. Vol. 1: Survey Method & Results*. Springfield: Illinois Department of Conservation, 1978.

Woods, Nicholas A. *Prince of Wales in Canada and the United States*. London: Bradbury and Evans, 1861.

Young, Dick. *Kane County Wild Plants and Natural Areas*. 2nd ed. Geneva, Ill.: Kane County Forest Preserve District, 1994.

Index to Species

Bold = primary reference

Avocet, American, 18, 58, 74, 162, 163, 164, **200**

Bittern: American, 17, 58, 129, 164, 169, **195**; Least, 9, 40, 51, 58, 61, 69, 76, 78, 81, 83, 88, 89, 95, 98, 129, 139, 149, 150, 151, 158, 168, **195**
Blackbird: Brewer's, 36, 51, 77, 103, 121, 153, 177, **213**; Red-winged, 6, 163, **213**; Rusty, 7, 13, 86, 89, 105, 157, 181, **213**; Yellow-headed, 11, 42, 43, 59, 60, 69, 78, 83, 97, 98, 102, 123, 124, 149, 156, 158, **213**
Bluebird, Eastern, 7, 13, 24, 25, 38, 40, 43, 47, 49, 76, 77, 88, 89, 107, 115, 141, 143, 157, 165, 169, 177, **208**
Bobolink, 9, 11, 21, 23, 38, 40, 41, 42, 43, 52, 54, 62, 72, 79, 82, 88, 89, 95, 100, 104, 107, 108, 109, 111, 114, 116, 121, 126, 134, 149, 151, 153, 176, **213**
Bobwhite, Northern, 9, 65, 129, 133, 134, 135, 146, 178, 179, 181, **199**
Brant, 68, **196**
Bufflehead, 33, 64, 83, 124, 133, 174, **198**
Bunting: Indigo, 24, 25, **213**; Snow, 6, 14, 21, 28, 76, 103, 106, 122, 124, 144, 149, 153, 166, 178, **212**

Canvasback, 6, 41, 64, 74, 81, 132, 156, 174, **197**
Cardinal, Northern, **212**
Catbird, Gray, **208**
Chat, Yellow-breasted, 10, 25, 41, 47, 49, 54, 64, 65, 76, 82, 100, 104, 105, 106, 113, 114, 122, 124, 128, 129, 130, 133, 139, 141, 144, 146, 147, 150, 158, 169, 172, 177, 178, 179, 186, **211**

Chickadee, Black-capped, **207**
Chuck-will's-widow, 10, 47, 172, **204**
Collared-Dove, Eurasian, **203**
Coot, American, 27, 28, 42, 48, 51, 57, 59, 60, 64, 72, 81, 85, 89, 157, **200**
Cormorant, Double-crested, 7, 9, 18, 40, 42, 48, 60, 109, 124, 137, 156, **195**
Cowbird, Brown-headed, **213**
Crane, Sandhill, 6, 7, 12, 13, 24, 41, 53, 69, 72, 82, 83, 86, 87, 88, 89, 90, 92, 95, 98, 102, 114, 115, 121, 123, 124, 152, 156, 158, 159, 163, 165, 172, 178, 180, 181, 182, 191, **200**
Creeper, Brown, 24, 69, 71, 122, 132, 137, 140, 182, **207**
Crossbill: Red, 14, 89, 109, 115, 168, 176, **214**; White-winged, 14, 35, 76, 89, 109, 115, 168, 193, **214**
Crow, American, **206**
Cuckoo: Black-billed, 24, 25, 31, 34, 40, 48, 71, 90, 105, 109, 133, 158, 176, **203**; Yellow-billed, 24, 25, 31, 34, 36, 40, 63, 71, 90, 105, 109, 133, 141, 158, 176, **204**

Dickcissel, 9, 11, 40, 43, 52, 65, 79, 95, 102, 104, 106, 109, 114, 115, 134, 145, 178, **213**
Dove: Mourning, **203**; Rock, **203**
Dowitcher: Long-billed, 59, 95, 145, 162, 163, **202**; Short-billed, 145, 153, 162, **202**
Duck (*see also* Mallard): American Black, 41, 51, 126, 132, **197**; Harlequin, 17, 30, 74, 154, 190, **198**; Ring-necked, 33, 51, 64, 82, 132, 133, 139, 174, 179, **197**; Ruddy, 33, 60, 81, 132, **198**; Wood, 36, 42, 50,

51, 69, 72, 89, 105, 115, 118, 120, 154, 164, **197**

ducks, 6, 7, 19, 22, 27, 35, 42, 44, 48, 58, 59, 63, 69, 78, 79, 83, 85, 89, 90, 98, 103, 107, 109, 112, 113, 114, 121, 124, 127, 137, 151, 153, 158, 164, 165, 166, 180, 182, 185, 186, 187, 193

Dunlin, 36, 111, **202**

Eagle: Bald, 5, 6, 13, 14, 28, 47, 48, 51, 77, 89, 90, 107, 126, 127, 149, 179, 181, **198**; Golden, 28, 77, 178, 181, **199**

Egret (*see also* Heron; Night-Heron): Cattle, 113, 135, 137, **196**; Great, 37, 40, 42, 58, 60, 103, 137, 139, 168, 196; Snowy, 58, 78, 113, **196**

egrets, 11, 12, 38, 129, 163

Eider, King, 17, 68, 118, 166, **198**

Falcon (*see also* Kestrel, American; Merlin): Peregrine, 12, 13, 18, 28, 77, 151, **199**

falcons, 12, 28, 77

Finch: House, **213**; Purple, 49, 68, 109, 143, **213**

Flicker, Northern, 24, 172, **205**

Flycatcher (*see also* Wood-Pewee): Acadian, 49, 64, 65, 69, 71, 82, 90, 92, 96, 112, 114, 126, 134, 140, 141, 144, 170, 172, 189, 190, 191, **205**; Alder, 9, 41, 49, 52, 82, 88, 89, 152, 159, 172, **205**; Great Crested, 24, 48, 96, **206**; Least, 134, 205, **206**; Olive-sided, 36, 48, 76, 113, 163, **205**; Willow, 9, 24, 31, 36, 38, 46, 49, 58, 95, 107, 124, 141, 152, 157, 172, **205**; Yellow-bellied, 34, 76, **205**

Gadwall, 49, 82, 83, 126, 133, 151, **197**

geese, 85, 89, 109, 153, 182, 185

Gnatcatcher, Blue-gray, 25, 33, 46, 48, 50, 55, 64, 100, 152, 165, 190, **208**

Godwit: Hudsonian, 58, 59, 145, 162, **201**; Marbled, 145, 163, **201**

Golden-Plover, American, 7, 11, 23, 95, 128, 129, 135, 145, 153, 166, 177, 178, **200**

Goldeneye, Common, 13, 17, 28, 81, 109, 124, 126, 128, 133, 151, 156, **198**

Goldfinch, American, 185, **214**

Goose (*see also* Brant): Canada, 86, 95, 103, **196**; Greater White-fronted, 6, 41, 81,

95, 103, 124, 149, 176, **196**; Ross's, 103, **196**; Snow, 95, 103, 149, 196

Goshawk, Northern, 54, 150, 168, **199**

Grackle, Common, 6, 163, **213**

Grebe: Eared, 60, 67, **195**; Horned, 41, 67, 69, 89, 97, 98, 111, 118, 149, 151, 174, 179, **195**; Pied-billed, 11, 27, 28, 40, 43, 47, 51, 58, 60, 61, 64, 69, 76, 81, 83, 88, 89, 98, 102, 126, 134, 139, **195**; Red-necked, 154, **195**; Western, 166, **195**

grebes, 18, 20, 42, 48, 49, 85, 103, 109, 156, 164, 165, 185, 190

Grosbeak: Blue, 11, 134, 146, 178, 179, **212**; Evening, 109, 140, 185, **214**; Pine, 168, 193, **213**; Rose-breasted, 24, 38, 40, 50, **212**

Grouse, Ruffed, 191, **199**

Gull (*see also* Kittiwake): Bonaparte's, 8, 14, 33, 75, 124, **203**; Franklin's, 13, 21, 63, 75, **202**; Glaucous, 17, 62, 72, 75, 118, 128, 151, 161, **203**; Great Black-backed, 17, 62, 128, 161, **203**; Herring, 17, 57, 62, **203**; Iceland, 17, 62, 78, 128, 151, **203**; Laughing, 75, 154, 190, **202**; Lesser Black-backed, 17, 62, 78, 118, 128, **203**; Little, 13, 14, 42, 154, 164, 190, **202**; Ring-billed, 10, 17, 33, 57, 62, 75, 128, 132, **203**; Sabine's, 28, 164, 185, **203**; Thayer's, 17, 62, 72, 128, **203**

gulls, 5, 6, 19, 23, 30, 31, 59, 77, 85, 89, 106, 109, 119, 126, 127, 139, 156, 162, 186, 188

Harrier, Northern, 5, 6, 13, 40, 52, 55, 77, 102, 103, 107, 114, 115, 124, 129, 134, 176, 180, **198**

Hawk (*see also* Falcon; Harrier; Osprey): Broad-winged, 7, 12, 24, 26, 38, 48, 50, 51, 54, 55, 69, 71, 77, 97, 106, 107, 111, 112, 118, 119, 122, 132, 140, 141, 144, 170, **199**; Cooper's, 25, 32, 43, 50, 77, 88, 100, 106, 107, 111, 140, 141, 150, 156, 157, **199**; Red-shouldered, 5, 6, 31, 36, 37, 43, 69, 77, 106, 111, 115, 126, 138, 140, 146, 153, 170, 177, 190, 191, **199**; Red-tailed, 77, 107, **199**; Rough-legged, 5, 14, 55, 77, 89, 95, 103, 107, 113, 122, 123, 129, 133, 134, 150, 176, 180, **199**; Sharp-shinned, 77, 106, 107, 150, **198**; Swainson's, 96, 116, 122, **199** hawks, 53, 63, 72, 75, 90, 149, 163, 178, 188

Heron (*see also* Egret; Night-Heron): Great Blue, 37, 42, 53, 58, 60, 83, 87, 90, 103, 137, 139, 172, **195**; Green, 28, 31, 37, 61, 154, **196**; Little Blue, 58, 78, 113, **196**; Tricolored, **196**

herons, 6, 9, 10, 11, 12, 27, 38, 40, 41, 57, 63, 69, 78, 79, 81, 82, 100, 102, 107, 108, 109, 112, 114, 115, 118, 119, 124, 126, 129, 132, 151, 162, 176, 179

Hummingbird, Ruby-throated, 27, 47, 50, 64, 96, 152, 156, 166, 191, **204**

Jaeger: Long-tailed, **202**; Parasitic, 166, **202**; Pomarine, **202**

jaegers, 12, 13, 18, 28, 154, 164

Jay, Blue, 172, **206**

Junco, Dark-eyed, 122, **212**

Kestrel, American, 24, **199**

Killdeer, 105, 133, **200**

Kingbird, Eastern, 27, 49, **206**

Kingfisher, Belted, 24, 31, 37, 50, 130, **205**

Kinglet: Golden-crowned, 24, 33, 68, 69, 76, 118, **208**; Ruby-crowned, 33, 34, **208**

Kittiwake, Black-legged, 13, 28, 164, 166, **203**

Knot, Red, 59, 74, 145, 154, **201**

Lark, Horned, 6, 14, 21, 23, 28, 83, 103, 122, 124, 143, 149, 153, 178, **207**

Longspur: Lapland, 6, 7, 14, 23, 28, 103, 106, 122, 124, 144, 149, 153, 166, 178, **207**; Smith's, 7, 178, **212**

Loon: Common, 7, 13, 41, 47, 48, 69, 85, 89, 90, 97, 98, 103, 149, 151, 174, **195**; Red-throated, 12, 28, 154, 164, 172, **195**

loons, 18, 20, 42, 49, 67, 156, 165, 185, 186

Mallard, 72, **197**

Martin, Purple, 8, 18, **207**

Meadowlark: Eastern, 31, 38, 40, 41, 54, 62, 88, 95, 105, 114, 133, 164, **213**; Western, 52, 65, 77, 95, 106, 116, 133, 134, 144, 145, 158, 177, **213**

Merganser: Common, 6, 33, 49, 59, 81, 124, 126, 156, **198**; Hooded, 27, 33, 42, 47, 49, 72, 81, 88, 118, 124, 133, 134, 146, 158, 179, **198**; Red-breasted, 17, 33, 49, 81, 124, **198**

mergansers, 13, 14, 28, 72, 97, 132, **151**

Merlin, 12, 13, 18, 28, 77, 107, 151, **199**

Mockingbird, Northern, 9, 134, 135, 145, 146, 178, 185, **208**

Moorhen, Common, 28, 43, 47, 51, 57, 58, 59, 60, 61, 69, 78, 81, 82, 83, 88, 95, 98, 108, 123, 124, 134, 137, **200**

Night-Heron: Black-crowned, 7, 21, 28, 31, 36, 40, 42, 43, 58, 68, 97, 103, 133, 137, 163, 164, 168, **196**; Yellow-crowned, 34, 61, 68, 69, 76, 139, 168, **196**

Nighthawk, Common, 12, 154, **204**

Nuthatch: Red-breasted, 76, 88, 98, 143, **207**; White-breasted, 37, **207**

Oldsquaw, 5, 13, 17, 20, 28, 65, 72, 128, 151, **198**

Oriole: Baltimore, 33, **213**; Orchard, 11, 36, 40, 41, 47, 49, 52, 61, 65, 71, 95, 106, 124, 126, 133, 134, 135, 139, 144, 146, 149, 151, 158, 164, 177, 178, 179, **213**

Osprey, 7, 12, 47, 48, 49, 77, 89, 98, 106, 107, 115, 119, 124, 135, 149, 179, **198**

Ovenbird, 38, 40, 43, 50, 64, 82, 96, 118, 124, 126, 132, 140, 141, 144, 188, 189, **210**

Owl (*see also* Screech-Owl): Barred, 50, 65, 69, 71, 122, 124, 126, 141, 144, 146, 150, 165, 170, 173, 177, 181, 182, 189, 191, **204**; Great Horned, 5, 8, 32, 51, 54, 55, 92, 109, 120, 141, 150, 173, **204**; Long-eared, 6, 38, 52, 55, 75, 90, 103, 113, 114, 115, 134, 141, 150, 153, **204**; Northern Saw-whet, 6, 38, 43, 63, 114, 115, 141, 150, **204**; Short-eared, 6, 7, 14, 21, 28, 40, 52, 55, 61, 95, 102, 113, 114, 129, 134, 143, 149, 150, 151, 161, 176, **204**; Snowy, 5, 14, 17, 18, 20, 21, 72, 74, 150, 151, 161, 166, 190, **204**

owls, 6, 14, 97, 109, 152, 157

Parakeet, Monk, 28, 57, **203**

Partridge, Gray, 153, **199**

Parula, Northern, 53, 170, 190, **209**

Pelican, American White, 42, 132, 137, 149, **195**

Pewee. *See* Wood-Pewee

Phalarope: Red, 13, 72, **202**; Red-necked, 145, 154, 163, **202**; Wilson's, 23, 40, 43, 59, 85, 95, 111, 113, 145, 176, **202**

Pheasant, Ring-necked, **199**

Phoebe, Eastern, 32, 50, 76, 144, **206**

Pigeon. *See* Rock Dove

Pintail, Northern, 6, 41, 49, 85, 118, 126, 132, **197**

Pipit, American, 8, 13, 28, 83, 95, 103, **208**

Plover (*see also* Golden-Plover, American; Killdeer): Black-bellied, 23, 36, 128, 145, 153, 162, **200**; Piping, 74, 154, **200**; Semipalmated, 43, 85, **200**

Rail (*see also* Sora): King, 59, 76, 134, 139, 149, 150, 162, **200**; Virginia, 7, 25, 42, 43, 49, 54, 58, 59, 61, 81, 83, 98, 100, 107, 119, 123, 138, 149, 150, 151, 161, 169, 186, **200**; Yellow, 17, 178, **200**

Redhead, 6, 64, 74, 81, 82, 97, 132, 174, **197**

Redpoll, Common, 35, 48, 76, 157, 168, 176, 186, **214**

Redstart, American, 41, 49, 51, 63, 82, 92, 106, 123, 133, 134, 144, 146, 166, 172, 176, **210**

Robin, American, 69, 109, **208**

Ruff, 60, **202**

Sanderling, 18, 74, **201**

Sandpiper (*see also* Ruff; Sanderling): Baird's, 43, 59, 145, 153, **201**; Buff-breasted, 11, 12, 18, 95, 128, 135, 145, 153, 166, 177, **202**; Least, 145, 153, **201**; Pectoral, 7, 145, 153, 162, **202**; Purple, 13, 21, 30, 72, 75, 154, 166, 190, **202**; Semipalmated, 145, **201**; Solitary, 145, 171, **201**; Spotted, 37, 111, 154, **201**; Stilt, 145, 163, **202**; Upland, 10, 11, 52, 55, 76, 95, 114, 133, 134, 135, 145, 149, 151, 153, 166, 177, 178, **201**; Western, 154, 162, **201**; White-rumped, 23, 43, 59, 111, **201**

Sapsucker, Yellow-bellied, 33, 38, **205**

Scaup: Greater, 28, 132, 151, **197**; Lesser, 133, **197**

scaup, 13, 179

Scoter: Black, 6, 13, 17, 154, 166, **198**; Surf, 6, 13, 17, 154, 166, **198**; White-winged, 6, 17, 128, 151, 154, 161, 166, **198**

scoters, 13, 28, 31, 67, 74, 78, 190

Screech-Owl, Eastern, 32, 38, 52, 92, 122, 141, 146, 150, 173, **204**

shorebirds, 7, 8, 9, 27, 28, 35, 36, 41, 42, 50, 51, 57, 58, 60, 68, 69, 77, 78, 79, 82, 83, 98, 102, 103, 104, 105, 107, 109, 113, 121, 123, 132, 136, 137, 152, 156, 176, 185, 188,

Shoveler, Northern, 49, 51, 58, 81, 83, 139, **197**

Shrike: Loggerhead, 9, 10, 114, 129, 134, 181, **206**; Northern, 6, 14, 31, 40, 42, 74, 76, 77, 89, 95, 103, 113, 123, 168, 176, **206**

Siskin, Pine, 35, 48, 76, 109, 120, **214**

Snipe, Common, 8, 26, 34, 40, 50, 78, 90, 105, 129, 134, 145, 151, 164, 186, **202**

Sora, 7, 25, 41, 42, 49, 54, 60, 61, 81, 97, 107, 119, 123, 138, 149, 150, 161, 186, **200**

Sparrow (*see also* Junco, Dark-eyed; Towhee, Eastern): American Tree, 122, 143, 163, **211**; Chipping, **211**; Clay-colored, 61, **211**; Field, 31, 40, 47, 49, 62, 121, 172, 178, **211**; Fox, 76, 105, **212**; Grasshopper, 11, 38, 40, 43, 52, 55, 69, 79, 82, 88, 95, 102, 104, 108, 109, 114, 115, 126, 134, 143, 145, 146, 149, 153, 164, 176, 178, 179, **211**; Harris's, 150, 161, **212**; Henslow's, 11, 43, 52, 55, 76, 95, 108, 114, 129, 133, 134, 147, 149, 164, 178, 179, **212**; House, **214**; Lark, 11, 134, 135, 146, 147, 178, 179, **211**; Le Conte's, 8, 26, 151, 161, **212**; Lincoln's, **212**; Nelson's Sharp-tailed, 12, 108, 114, 151, 178, **212**; Savannah, 8, 40, 42, 43, 72, 107, 108, 114, 116, 121, 161, 178, **211**; Song, 8, **212**; Swamp, 8, 54, 58, 138, 151, 157, 158, 169, 172, **212**; Vesper, 8, 10, 40, 42, 69, 76, 79, 133, 145, 147, 149, 161, **211**; White-crowned, 186, **212**; White-throated, 34, **212**

sparrows, 13, 17, 20, 23, 26, 27, 28, 35, 36, 65, 77, 90, 92, 118, 119, 123, 158, 163, 171, 180, 186

Starling, European, **208**

Swallow (*see also* Martin, Purple): Bank, 21, 41, 58, 64, 185, **207**; Barn, 58, 72, **207**; Cliff, 49, 58, 130, 135, **207**; Northern Rough-winged, 37, 54, 58, 126, **207**; Tree, 8, 25, 58, **207**
swallows, 33, 47, 51, 153, 172
Swan: Mute, 5, 59, **197**; Tundra, 6, 13, 42, 51, 59, 78, 85, 89, 95, 121, 179, **197**
Swift, Chimney, 27, **204**

Tanager: Scarlet, 8, 34, 37, 38, 40, 46, 47, 48, 50, 51, 54, 55, 96, 109, 118, 120, 124, 126, 141, 144, 146, 152, 156, 158, 164, **211**; Summer, 10, 24, 34, 133, 134, 147, 178, 179, 181, **211**
Teal: Blue-winged, 11, 36, 49, 58, 60, 64, 81, 132, 139, **197**; Green-winged, 49, 82, 109, 126, 132, 139, **197**
Tern: Black, 9, 42, 43, 82, 83, 87, 98, 102, 113, 126, 149, 156, 157, 158, 174, **203**; Caspian, 12, 75, 97, **203**; Common, 8, 75, **203**; Forster's, 8, 9, 79, 83, 86, 87, 90, 92, 97, 113, 149, 152, 158, **203**
terns, 18, 59, 89, 103, 123, 127, 137, 139, 151, 162, 164, 185, 188, 190
Thrasher, Brown, 31, 33, 63, **208**
Thrush (*see also* Robin, American; Veery): Gray-cheeked, 32, **208**; Hermit, 65, 68, 69, 76, 118, **208**; Swainson's, 32, 34, **208**; Varied, **208**; Wood, 41, 43, 47, 48, 50, 51, 64, 82, 105, 118, 120, 124, 140, 141, 144, 145, 152, 165, 166, 182, **208**
thrushes, 8, 20, 24, 31, 33, 36, 37, 59, 63, 75, 90, 92, 96, 106, 122, 123, 126, 127, 146, 158, 171
Titmouse, Tufted, 24, 38, 44, 46, 47, 48, 50, 51, 63, 64, 65, 70, 111, 118, 120, 122, 124, 126, 133, 141, 143, 145, 146, 163, 181, 189, **207**
Towhee, Eastern, 8, 122, 164, **211**
Turkey, Wild, 158, **199**
Turnstone, Ruddy, 74, 145, **201**

Veery, 9, 25, 32, 40, 41, 49, 50, 63, 64, 69, 71, 89, 96, 120, 140, 141, 146, 170, 182, **208**
Vireo: Bell's, 10, 42, 48, 52, 59, 65, 104, 114, 128, 129, 130, 133, 134, 139, 141, 144, 146, 177, 178, 179, 181, 182, **206**; Blue-headed, **206**; Philadelphia, 33,
112, **206**; Red-eyed, 27, 96, 109, 118, 141, **206**; Warbling, 27, 31, 58, 102, 105, 109, 121, 124, 141, **206**; White-eyed, 41, 46, 54, 63, 65, 88, 89, 105, 106, 112, 114, 122, 128, 130, 133, 139, 141, 144, 158, 169, 170, 172, 186, 187, **206**; Yellow-throated, 38, 49, 55, 63, 82, 92, 106, 109, 118, 126, 141, 156, 158, 189, **206**
vireos, 24, 34, 36, 37, 38, 40, 44, 68, 69, 75, 90, 119, 120, 123, 126, 140
Vulture, Turkey, 7, 48, 96, 107, 115, 132, 144, **196**

Warbler (*see also* Chat, Yellow-breasted; Ovenbird; Parula, Northern; Redstart, American; Yellowthroat, Common): Bay-breasted, **210**; Black-and-white, 71, 140, 176, 187, 188, **210**; Black-throated Blue, 53, 111, **209**; Black-throated Green, 10, 187, 188, **209**; Blackburnian, **209**; Blackpoll, 68, **210**; Blue-winged, 32, 38, 40, 46, 47, 48, 49, 65, 71, 82, 88, 96, 108, 115, 122, 141, 150, 152, 157, 158, 170, 173, 188, 191, **209**; Canada, 32, 33, 71, 77, 120, 170, 187, **211**; Cape May, **209**; Cerulean, 10, 43, 49, 69, 71, 92, 106, 111, 113, 134, 141, 144, 170, 172, 176, 182, 189, 190, **210**; Chestnut-sided, 41, 54, 76, 90, 152, 173, 176, 188, **209**; Connecticut, 18, 32, 33, 36, 54, 68, 76, 100, 106, 111, 112, 113, 114, 120, 123, 141, 161, 173, **210**; Golden-winged, 38, 71, 76, 141, 158, 173, **209**; Hooded, 10, 26, 33, 43, 47, 50, 51, 54, 69, 76, 106, 111, 141, 154, 159, 170, 172, 187, 188, 189, **211**; Kentucky, 10, 32, 54, 64, 65, 69, 106, 132, 134, 140, 141, 144, 154, **210**; Magnolia, **209**; Mourning, 18, 32, 33, 36, 54, 76, 77, 88, 112, 114, 120, 134, 152, 173, **210**; Nashville, 159, **209**; Orange-crowned, **209**; Palm, **210**; Pine, 32, 53, 88, **210**; Prairie, 10, 170, 172, **210**; Prothonotary, 10, 24, 26, 32, 44, 98, 106, 111, 112, 113, 120, 132, 133, 137, 144, 145, 146, 154, 165, 170, 176, 177, 182, 186, **210**; Tennessee, **209**; Wilson's, 33, **211**; Worm-eating, 10, 26, 48, 68, 88, 106, 173, 187, **210**; Yellow, 27, 28, 46, 49, 89, 157, 158, **209**; Yellow-rumped, 24, 34, 69, 106, 118, 119, **209**; Yellow-throated, 10, 166, 172, 182, 190, **209**

warblers, 5, 8, 20, 31, 35, 37, 42, 55, 57, 58, 59, 60, 61, 63, 67, 75, 81, 83, 86, 95, 97, 102, 105, 107, 109, 121, 126, 127, 138, 143, 149, 153, 163

Waterthrush: Louisiana, 8, 10, 26, 48, 50, 65, 68, 106, 140, 141, 146, 170, 189, 190, 191, **210**; Northern, 152, **210**

Waxwing: Bohemian, 109, **208**; Cedar, 109, 185, **209**

Whimbrel, 18, 74, 145, 154, 164, **201**

Whip-poor-will, 10, 47, 76, 78, 90, 133, 134, 146, 158, 161, 179, 181, 188, **204**

Wigeon: American, 49, 72, 83, 85, 126, 132, 133, 151, **197**; Eurasian, 49, **197**

Willet, 8, 58, 74, 85, **201**

Wood-Pewee, Eastern, 37, 48, 96, 127, **205**

Woodcock, American, 7, 25, 26, 38, 40, 54, 58, 61, 64, 78, 81, 100, 105, 109, 111, 113, 114, 124, 129, 141, 144, 161, 163, 164, 169, 180, 186, **202**

Woodpecker (*see also* Flicker, Northern; Sapsucker, Yellow-bellied): Downy, 24, **205**; Hairy, 24, **205**; Pileated, 24, 44, 70, 146, 166, 177, 181, 182, 183, 188, 189, 191, **205**; Red-bellied, 24, 141, **205**; Red-headed, 24, 26, 32, 51, 63, 64, 98, 106, 115, 143, 144, 146, 156, 170, 177, **205**

Wren: Carolina, 44, 48, 51, 63, 64, 65, 111, 123, 126, 139, 141, 146, 164, 181, **207**; House, **207**; Marsh, 40, 47, 58, 81, 89, 105, 111, 114, 150, 152, 169, **208**; Sedge, 9, 10, 38, 40, 42, 43, 52, 55, 78, 89, 95, 100, 104, 105, 108, 114, 138, 141, 149, 150, 152, 157, 158, 164, 169, 176, **207**; Winter, 8, 34, 50, 65, 69, 111, 119, 186, 189, **207**

Yellowlegs: Greater, 7, 43, 85, 145, **200**; Lesser, 7, 36, 43, 85, 145, **201**

Yellowthroat, Common, 27, 28, 49, 121, 172, **211**

Index to Sites

Bold = pages with maps

Almond Marsh, **80**, 83
Arends Forest Preserve, Les, **117**, 120
Aroma Forest Preserve, **142**, 145

Bachelor Grove, **53**, 54
Baker's Lake Nature Preserve, **39**, 42
Baldy, Mt., **167**, 172
Bartel Grasslands, **53**, 55
Beaubien Woods Forest Preserve, **53**, **56**, 62
Beaver Lake Prairie Chicken Refuge, **175**, 178
Belmont Harbor, **163**, 19
Bemis Woods, **16**, 25
Benton Harbor, **184**, 185
Bergman Slough, **46**, 48
Beulah Bog, 159
Beverly Shores, **167**, **171**, 172
Big Foot Beach State Park, **156**, 157
Big Marsh, **53**, **56**, 58
Big Rock Forest Preserve, **117**, 124
Black Partridge Forest Preserve, 44, **45**
Black Tern Marsh, 98, **99**
Blackhawk Forest Preserve, **117**, 119
Blackwell Forest Preserve, **101**, 107
Bliss Woods Forest Preserve, **117**, 123
Bloomfield State Wildlife Area, **156**, 157
Blue Island Cemeteries, 52, **53**
Bluhm Woods County Park, **167**, 174
Bode Lakes, **39**, 40
Bong State Recreation Area, **148**, 149
Bowen Park, **73**, **74**, 75
Braidwood Dunes & Savanna Nature Preserve, **131**, 134
Braidwood Fish & Wildlife Area, **131**, 135
Brandon Lock and Dam, **136**, 139
Brewster Creek Marsh, **101**, 102
Brookfield Zoo, 26

Brown Forest Preserve, Ned, 38, **39**
Brown Sanctuary, **184**, 186
Buffalo Creek Forest Preserve, 79, **80**
Buffalo Creek Nature Preserve, 81
Burr Oak Woods, **53**, 54
Burnham Harbor, **16**, 21
Burnham Prairie, **53**, **56**, 61
Burnidge Forest Preserve, **117**, 121
Busse Forest Nature Preserve, 38, **39**

Calumet College Marsh, **160**, 162
Calumet Park, **53**, **56**, 57
Calumet Sewage Treatment Plant, **53**, **56**, 60
Camp Lake, **148**, 150
Camp Logan, **76**, 78
Camp Sagawau, **45**, **46**, 47
Campbell Airport Area, **80**, 83
Cap Sauer's Holdings Nature Preserve, **45**, **46**, 48
Cedar Lake, Ill., **84**, 87
Cedar Lake, Ind., 165, **175**
Chain O'Lakes State Park, **84**, 90, **91**
Channahon State Park, **128**, 130, **131**
Cherry Hill Woods, **45**, **46**, 49
Chicago Botanic Garden, **29**, **35**, 35
Chicago Portage Woods, **16**, 25
Chicago River North Branch, 26, 32, 69
Chipilly Woods Forest Preserve, **29**, 32
Chiwaukee Prairie, **148**, 151
Cliffside Park, **148**, 153
Colonial Park, **148**, 154
Columbia Woods, 44, **45**, **46**
Columbus Park, **16**, 27
Como Lake, **156**, 157
Comus Lake, **156**, 158
Conkey Forest Preserve, **53**, 55
Coral Woods, **93**, 96
Cowles Bog, **167**, 169
Crabtree Nature Center, **39**, 41

Creek Ridge County Park, **167**, 174
Crow Island Park, **34**, 34
Crystal Lake, **93**, 97
Cuba Marsh, **80**, 81

Dam No. 1, **29**, 38
Danada Forest Preserve, **101**, 109
Daniel Wright Woods, **66**, 71
Deadstick Pond, **53**, **56**, 58
Deep Lake, **84**, 87
Deep River County Park, **160**, 165
Deer Grove Forest Preserve, **39**, 42
Delavan Inlet, **156**, 158
Denny Road Marsh, **117**, 124
Des Plaines Conservation Area, **128**, **131**, 133
Des Plaines Wetlands Project, **73**, 78
DeTonty Woods, **46**
Diehl Road Wetlands, 113
Douglas Forest Preserve, Paul, **39**, 43
Douglas Nature Sanctuary, Paul, 21, **22**
Dresden Lock and Dam, **128**
Dresden Ponds, **128**, **131**, 132
Druce Lake, **84**, 85
Drury Lane Pothole, **84**, 85
Duck Lake Nature Area, **156**, 158
Duffy Preserve, John J., **45**, **46**, 49
DuPage Airport, **101**, 101

Eagle Point on Pistakee Lake, **84**, 90
East Loon and West Loon Lake, **84**, 87
Eggers Woods Forest Preserve, **53**, **56**, 59
83rd St. Marsh, 114
Elburn Forest Preserve, **117**, 122
Elizabeth Conkey Forest Preserve, **53**, 55
Elsen's Hill, **101**, 106
Emily Oaks Nature Center, 219
Eola Road Marsh, **101**, 113
Erickson Woods, **29**, **34**, 33
Evanston Art Center, 30
Evanston Ecology Center, 219

Fabyan Forest Preserve, **117**, 120
Fermilab, **101**, 103, **104**
Fernwood Botanic Garden and Nature Center, **184**, 191
Ferson Creek Fen Nature Preserve, **117**, 119
Flatfoot Lake, **53**, **56**, 62
Floral Road, **184**, **188**, 188

Foley Park, **66**, 68
Forest Lawn and Lakeside Roads, **184**, 190
Forest Park and Beach. *See* Lake Forest Beach
Forsythe Park, **56**, 162
Fort Sheridan, **66**, 67
Four Seasons Nature Center, 155
Fourth Lake, **84**, 86
Fox Lake, **84**, 90
Fox River Park, **148**, 150
Fullersburg Woods, **101**, 112
Fullerton Park, **101**, 103

Gage's Lake, **84**, 85
Gar Creek Trail and Prairie, **131**, 144
Gavin Prairie, 88
Gebhard Woods State Park, **125**, 127
Gibson Woods Nature Preserve, **148**, 163
Gillson Park, **29**, 30
Glacial Park Conservation Area, **93**, **94**
Glenview Woods, **29**, 30
Glenwood Park Forest Preserve, **117**, 120
Goodenow Grove Preserve, **142**, 143
Goose Lake Prairie State Park, **125**, **128**, 129
Grand Kankakee Marsh County Park, 165, **175**
Grand Mere State Park, **184**, 187
Grant Creek Prairie, **131**, 133
Grant Park, **16**, 20
Grant Woods Forest Preserve, **84**, 88
Grass Lake, **84**, 90
Great Lakes Naval Training Center, **66**, 68
Green Lake Woods, **53**, 63
Greene Valley Forest Preserve, **101**, 114
Grove, The, **29**, 36
Gurnee Mills Ponds, **73**, 79

Haf's Pond, **156**, 157
Half Day Forest Preserve, **66**, 71
Hammond Cinder Flats, **56**, **160**, 162
Hammond Lakefront Park, **56**, **160**, 161
Hampshire Forest Preserve, 116, **117**
Harms Woods, **29**, 32
Harris Forest Preserve, **125**, 127
Harrison-Benwell Conservation Area, 92, **93**
Hawthorne Hollow Arboretum, 150
Hegewisch Marsh, **53**, **56**, 60
Heidecke Lake State Fish and Wildlife Area, **125**, 127, **128**

Heller Nature Center, 220
Henry DeTonty Woods, **46**
Heron Pond, **56**
Heron Rookery, **167**, 172
Herrick Lake Forest Preserve, **101**, 108
Hickory Creek Preserve, **136**, 141
Hickory Grove Highlands, **93**, 99
Hidden Lake Forest Preserve, **101**, **110**, 111
Higginbotham Woods, **136**, 140
Hooper Branch Savanna, **142**, 147, 180
Hoosier Prairie State Nature Preserve, **160**, 165
Hoy, Mt., **101**, 107
Hudson Lake, 174, **175**
Humboldt Park, **16**, 27

Illinois & Michigan (I & M) Canal & Tow-path, 44, **45**, 127, 132
Illinois Beach State Park North Unit, **73**, **76**, 77
Illinois Beach State Park South Unit, **73**, **76**
Illinois Prairie Path, **101**, 102
Independence Grove, **66**, 72
Indian Ridge Marsh, **53**, **56**, 58
Indiana Dunes Heron Rookery, **167**, 172
Indiana Dunes National Lakeshore, **167**, 168, **171**
Indiana Dunes State Park, **167**, 170, **171**
Inland Marsh, **167**, 169
Iroquois County State Wildlife Area, **142**, 147, 180
Iroquois Woods Natural Area, **131**, 144
Isle a la Cache, **131**, 137
Ivanhoe Marsh State Wildlife Area, 155, **156**

Jackson Park, **16**, 21, **22**
James Park, **29**, 31
Jasper-Pulaski Fish and Wildlife Area, **175**, 180, **181**
Jean Klock Park, **184**, 185
Jeorse Park, **160**, 162
John J. Duffy Preserve, **45**, **46**, 49
Johnson Beach, **167**, 170, **171**
Johnsons Mound Forest Preserve, **117**, 122
Joliet Arsenal, **128**, **131**, 133
Jurgensen Woods Nature Preserve, **53**, 64

Kankakee Fish and Wildlife Area, **175**, 182
Kankakee River State Park, **131**, 144

Keepataw Preserve, **136**, 138
Kemil Beach/Dune Ridge Trail, **167**, **171**, 172
Kemper Lakes, **80**, 81
Kenosha Harbor, **148**, 151
Kettle Moraine State Forest, **156**, 158
Kingsbury Fish and Wildlife Area, **175**, **176**
Kishwauketoe Nature Preserve, **156**, 157
Klock, Jean Park, **184**, 185

LaPorte Lakes, 174, **175**
LaBagh Woods, **16**, 26
Lake Calumet, **53**, **56**, 57
Lake Como, **156**, 157
Lake Comus, **156**, 158
Lake Forest Beach, **66**, 67
Lake Forest Nature Preserve, **66**, 67
Lake Forest Open Lands, **66**, 68
Lake Geneva, 155, **156**
Lake George, **56**, **160**, 162
Lake Katherine Nature Preserve, **53**, 220
Lake La Grange, **156**, 158
Lake Michigan Migrant Trap, **56**, **160**, 161
Lake Renwick Heron Rookery Nature Preserve, **136**, 137
Lakewood Forest Preserve, **80**, 82
Larsen Park, **73**, 75
LaSalle Fish and Game Area, **175**, 177
LeRoy Oakes Forest Preserve, **117**, 122
Les Arends Forest Preserve, **117**, 120
Lincoln Park Bird Sanctuary, **16**, 18
Lincoln Park Zoo, **16**, 20
Lippold Park, **93**, 97
Little Calumet River Trail, **167**, 169
Little Red Schoolhouse Nature Center, **45**, **46**, 47
Lockport Prairie Nature Reserve, **136**, 139
Logan, Camp, **73**, 78
Long John Slough, **45**, **46**, 47
Long Lake, Ill., **84**, 86
Long Lake, Ind., **167**, 168
Loon Lake, **84**, 87
Lost Valley Marsh, **94**
Love Creek County Park and Nature Center, **184**, 191
Lulu Lake, **156**, 159
Ly-co-ki-we Horse and Hiking Trail, **167**, **171**
Lyman Woods, **101**, 112

Lyons Prairie & Marsh, **93**, 100
Lyons Woods, **73**, 75

MacArthur Woods Nature Preserve, **66**, 71
McClaughry Springs Woods, **45**, **46**, 50
McCormick Place, **16**, 21
McCormick Ravine, **66**, 67
McDonald Woods, Margaret Mix , **35**, 36
MacDonald Woods, **84**, 88
McDowell Grove, **101**, 113
McGinnis Slough, **45**, **46**, 51
McHenry County Sod Farms, **93**, 95
McHenry Dam, 98, **99**
McKee Marsh, **101**, 107
McKinley Woods Forest Preserve, **128**,
    131, 132
Magic Hedge, **16**, 18, **19**
Maple Grove Forest Preserve, **101**, 111
Maple Lake, **45**, **46**, 47
Marengo Ridge, **93**, 95
Marquette Park, **160**, 164
Meigs Field, **16**, 21
Messenger Woods Nature Preserve, **136**,
    140
Metropolitan Sanitary District, **53**, **56**, 60
Michigan City Harbor, **167**, 173
Midewin National Tallgrass Prairie, **128**,
    131, 133
Migrant Trap, **56**, **160**, 161
Millard Park, **66**, 67
Miller Beach, **160**, 164
Miltmore Lake, **84**, 86
Momence Sod Farms, **142**, 145
Momence Wetlands Nature Preserve, **142**,
    146
Monee Reservoir, **142**, 143
Monroe Harbor, **16**, 21
Montrose Harbor, **16**, 18, **19**
Montrose Point, **16**, 18, **19**
Moraine Hills State Park, **93**, 98, **99**
Moraine Park, **66**, 67
Moraine State Nature Preserve, **167**, 173
Morton Arboretum, **101**, 109, **110**
Mt. Hope Cemetery, 52, **53**
Mud Lake Bog Nature Preserve, **184**, 191

Naperville Polo Club, 135, **136**
Natureland County Park, **156**, 159
Navy Pier, **16**, 20
Ned Brown Forest Preserve, 38, **39**

Nelson Lake Marsh Nature Preserve, **117**,
    123
New Buffalo Harbor, **184**, 190
New Munster Wildlife Area, **148**, 150
Newton County Fields, **175**, 178
Nicholson Pond, **148**, 153
Nippersink Lake, **84**, 90
Nippersink Marsh, 83, **84**
Norris Woods Nature Preserve, **117**, 120
North Branch Chicago River, 26, 32, 69
North Park Village Nature Center, 219
North Point Marina, **73**, **76**, 78
North Shore Channel, **29**, 31
Northwestern University, 28, **29**

Oak Point Picnic Area, **84**, **91**, 92
Oak Ridge Prairie County Park, **160**, 164
Oakes Forest Preserve, LeRoy, **117**, 122
O'Brien Lock & Dam, **53**, **56**, 63
O'Hara Woods Nature Preserve, **136**, 137
O'Hare Post Office Ponds, **16**, 23
Old Plank Road Trail, **136**, 141
Old School Forest Preserve, **66**, 72
Olive Park, **16**, 20
127th St. SEPA Station **53**, 62
Orland Hills Forest Preserve, **45**, 52
Otter Creek Bend Wetland Park, **117**, 121
Ottowa Trail Woods Forest Preserve, **16**, 24

Palatine Road Marsh, **39**, 41
Palos Park Woods Forest Preserve, **45**, **46**,
    51
Palos West Slough, **45**, **46**, 50
Park Avenue, Glencoe, **29**, 31
Park Avenue, Highland Park, **66**, 67
Paul Douglas Forest Preserve, **39**, 43
Paul Douglas Nature Sanctuary, 21, **22**
Paul Wolff Forest Preserve, **117**, 121
Paw Paw Lake, **184**, 186
Paw Paw Woods Nature Preserve, **45**, **46**,
    47
Pembroke Township, **142**, 146
Perkins Woods, **29**, 31
Petrifying Springs Park, **148**, 150
Pilcher Park, **136**, 140
Pistakee Lake, **84**, 90
Plum Creek Forest Preserve, **53**, 65
Plum Creek Nature Center, **142**, 143
Poplar Creek Forest Preserve, **39**, 40
Port of Indiana, 166, **167**

Portwine Road, **29**, 38
Powderhorn Lake Forest Preserve, **53**, **56**, 61
Prairie Spring Park, **148**, 151
Prairie Wolf Slough, **66**, 69
Pratt's Wayne Woods, **101**, 102

Raccoon Grove, **142**, 143
Racine Lakefront, **148**, 155
Radio Tower Marsh, 113
Rainbow Beach, **16**, 23
Red Oak Nature Center, 221
Red-Wing Slough, 83, **84**
Reed-Turner Woodland, **80**, 81
Renwick, Lake, **136**, 137
Rice Lake, **101**, 109
River Bend Nature Center, 221
River Trail Nature Center, **29**, 37
Riverdale Quarry, **53**, 60
Riverview Park, **184**, 185
Rochester Area, **148**, 152
Romeoville Prairie Nature Preserve, **136**, 138
Root River, **148**, 154, 156
Rosehill Cemetery, **16**, 27
Rosewood Beach, 65, **66**
Round Lake, **84**, 86
Roxana Pond, **56**, **160**, 163
Rubio Woods, **53**, 55
Rush Creek Conservation Area, 92, **93**
Rutland Forest Preserve, 116, **117**
Ryders Woods, **93**, 96
Ryerson Conservation Area, **66**, **70**

Saganashkee Slough, **45**, **46**, 48
Sagawau, Camp, **45**, **46**, 47
St. Francis Woods, **66**, 72
St. Joseph/Benton Harbor, **184**, 185
St. Margaret Mercy Health Care Center, 65, **160**
Salt Creek Woods Nature Preserve, **16**, 25
Sand Pond, **76**, 78
Sand Ridge Nature Center, **53**, 63
Sand Ridge Savanna Nature Preserve, **131**, 135
Sarett Nature Center, **184**, 186
Sauer's Holdings, Cap, **45**, 48
Sauganash Prairie, **16**, 26
Saw Wee Kee Park, 126
Schneider Sod Farms, 166, **175**

79th Street Beach, **16**, 23
Shamrock Turf Nurseries, **175**, 177
Shaw Woods and Prairie, **66**, 68
Shedd Aquarium, **16**, 21
Sheridan, Fort, **66**, 67
Shoe Factory Road Prairie, **39**, 40
Silver Springs State Park, 124, **125**
Skokie Lagoons, **29**, 33, **34**
Sod Farms, Indiana, 166, **175**, 177
Sod Farms, Kankakee County, **142**, 145
Sod Farms, McHenry County, **93**, 95
Sod Farms, Racine County, **148**, 153
Sod Farms, Will County, 135, **136**
Somme Prairie, **29**, 32
Songbird Slough Forest Preserve, **101**, 102
Spears Woods Forest Preserve, **45**, **46**, 46
Spring Bluff Forest Preserve, **73**, **76**, 78
Spring Brook Nature Center, **101**, 220
Spring Creek Valley Forest Preserve, **39**, 40
Spring Lake Nature Preserve, **39**, 40
Spring Valley Nature Center, 219
Springbrook Prairie Forest Preserve, **101**, 113
State Line Beach, **56**, **160**, 161
Sterling Lake, **73**, 79
Sterne's Woods Nature Preserve, **93**, 97
Stickney Run, **93**, 98
Stillman Nature Center, 219
Sugar Grove Marsh, **117**, 124
Summerhill Estates, **80**, 82
Sun Lake, **84**, 87
Sunset Hill Farm County Park, **167**, 173
Swallow Cliff Woods, **45**, **46**, 50

Tampier Lake, **45**, **46**, 49
Tekakwitha Woods, **117**, 119
Thatcher Woods, **16**, 24
Third Lake, **84**, 86
Thorn Creek Nature Preserve, 141, **142**
Thornton-Lansing Road Nature Preserve, **53**, 64
Tichigan Marsh, **148**, 152
Timber Ridge Forest Preserve, **101**, 106
Tinley Creek Woods Forest Preserve, **53**, 54
Tiscornia Beach Park, **184**, 185
Torrence Avenue SEPA, **53**, **56**, 62
Tower Road Park, **29**, 30
Townline Pond, 155, **156**
Trailside Museum, **16**, 24
Trout Park Nature Preserve, **117**, 118
Turnbull Woods, 36

Turtlehead Lake, **53**, 54
Tyler Creek Forest Preserve, **117**, 121

Valparaiso Lakes, **167**, 173
Van Patten Woods, **73**, 79
Van Vlissingen Prairie, **53**, **56**, 57
Veteran Acres Park, 97
Veteran's Woods, **136**, 138
Virgil Gilman Trail, **117**, 123
Vollmer Road Wildlife Marsh, **53**, 55
Volo Bog State Natural Area, **84**, 89

Wadsworth Prairie, **73**, 79
Wadsworth Wetlands Demonstration Project, **73**, 78
Wampum Lake, **53**, 64
Warbler Walkway, **156**, 158
Warren Dunes State Park, **184**, **188**
Warren Woods State Park, **184**, 189
Washington Park, **16**, 28
Waterfall Glen Forest Preserve, **101**, 115
Waverly Beach, **167**, 170, **171**
Waukegan Beach, **73**, **74**, 74
Waukegan Harbor, 72, **73**, **74**
West Beach, **167**, 168
West Chicago Prairie, **101**, 105
West DuPage Woods, **101**, 106

West Loon Lake, **84**, 87
Wetlands Demonstration Project, **73**, **78**
Whihala Beach County Park, **56**, **160**, 162
Whiting Park, **56**, **160**, 161
Widewaters, **128**, **131**, 132
Will-Cook Pond, **45**, **46**, 49
Will County Sod Farms, 135, **136**
William W. Powers Conservation Area, **53**, **56**, 59
Willow Slough Fish and Wildlife Area, **175**, 178, **179**
Wind Lake Sod Farms, **148**, 152
Wind Point, **148**, 154
Winnetka Avenue Bridge, **29**, **34**, 33
Wolf Lake, Ill., **53**, **56**, 59
Wolf Lake, Ind., **56**, **160**, 162
Wolf Lake Sanctuary, **53**, **56**, 59
Wolf Road Prairie Nature Preserve, **16**, 25
Wolff Forest Preserve, Paul, **117**, 121
Wooded Island, 21, **22**
Wooster Lake, **84**, 86
Wright Woods Forest Preserve, **66**, 71

Yellow Birch Trail, **184**, **188**, 188
Yorkville Dam, **125**, 126

Zander Woods, **53**, 64